SPACE - Ever Farther, Ever Faster - NOW!

FROM CRADLE TO EXOPLANETS

A CALL TO ACTION

By

Ed Gibson

Science Pilot

Skylab III

DRIVEN BY OUR HUMAN SPIRIT, CAPABILITIES
OF EARTH'S PEOPLE, AND THE VISION,
COURAGE, AND COMMITMENT OF
OUR LEADERS,

WE HUMANS HAVE BEGUN!

This book originated from public presentations made at the Astronaut Encounter at the NASA Kennedy Space Center Visitor Complex at the Kennedy Launch Facilities in Florida, sponsored by the Delaware North Corporation. The experience of making presentations on many of the subjects in this book to audiences composed of 50 to 200 visitors to the Center, three times a day for a week, once or twice a year, and over ten years has provided me with a wealth of material to consider for discussion. I thank Jon McBride and Nick Thomas for that opportunity. Although the scope of many subjects has been significantly amplified, some of the original content remains.

ISBN: 978-1-965384-49-7

© 2019 by Edward G. Gibson

No part of this book may be reproduced by any mechanical, photographic, electronic process or phonographic recording, nor may it be stored in a retrieval system, transmitted, or otherwise copied for public or private use, without written permission from the publisher.

Contents

Acknowledgments vi
Foreword ix
Scope xi
Dedication xiii
Chapter 1 America's Initial Impulse 1
Chapter 2 Another Low for America 4
Chapter 3 America's Golden Era in Space Begins 8
Chapter 4 What President Kennedy Left Behind 12
Chapter 5 Mercury – Making It Happen 15
Chapter 6 Gemini – A Steppingstone to Greater Spaceflight Capabilities 19
Chapter 7 Apollo – The Pinnacle 27
Chapter 8 Apollo 7-13 Apollo 7 – First Manned Flight of Apollo 37
   Apollo 8 39
   AMERICAN ENTHUSIASM RAN HIGH! 46
   Apollo 13 – NASA's Finest Hour 51
Chapter 9 Each Mission Improves Over the One Before 55
   Apollo 14 55
   Al Shepard Finally Flies Again 55
   Apollo 15 57
   The First of the New Apollo Capabilities 57
   Galileo's Theory Tested on the Moon 58
   Apollo 16 60
   Even Better Yet 60
   Apollo 17 A Scientist with Operational Skills and Geological Knowledge Explores the Moon 65
   Apollos 18-20 71
   Promise of the Most Productive Missions Yet! 71
Chapter 10 Payoffs from Apollo 73
   America's immediate payoffs came right down here, at home — in AMERICA! 74

Chapter 11 79

Our Depressing Plunge 79

   SUMMARY OF AMERICA IN SPACE 80

Chapter 12 Skylab 85

   Skylab I 90

   Skylab II 93

   Skylab III 96

Chapter 13: Apollo Soyuz - The first Joint Mission with the Soviets 130

Chapter 14   The Space Shuttle 134

   SPHEREx 143

II – Our First Major International Cooperation 151

Chapter 15 International Space Station 151

   Science 154

   Americans in Space 161

Chapter 16 America Struggles in Space 163

   Our Decreased Launch Capability 166

III – Cost Reduction 169

Chapter 17. Lowering Cost 169

IV —Where to First 190

Chapter 18 Moon or Mars? 190

Chapter 19 *Why?* 195

   ARTEMIS 209

Chapter 20 Moon First 214

   Our Solar System – Inner Planets 214

   MERCURY 215

   VENUS 216

   EARTH 217

   MARS 218

Chapter 21 Our Solar System – Outer Planets 231

   EUROPA 237

   TITAN 247

V HUMAN EXPLORATION EXPANDS 256

Chapter 22 256

Beyond Our Solar System to 256

EXOPLANETS! 256

Chapter 23 Vision, Courage, and Commitment Where Are We Today? 261

Chapter 24. Our Drive to Explore 265

Chapter 25   Where Does Human Go Now? 269

APPENDIX 271

    Abbreviations 271

Book References 274

Skylab References 275

Further Considerations on The Lunar Gateway 277

Ed Gibson Resume – Brief Summary 283

Ed Gibson Resume – Briefer Summary 284

We All Are Fortunate! 285

Focus on Students 287

Note to Author… 289

Ed Gibson

Acknowledgments

With a broad smile on my face, I thank Julie, my best friend for 70 years, my wife for 65 years, and whose support and patience have helped make this book possible in so many ways.

I also thank another Julie, my daughter, whose computer expertise, and enthusiastic support has also helped to improve the result significantly.

My appreciation goes also to my son, John, who's skill, unyielding drive to succeed, and great humor has been an inspiration to me ever since I watched him as just a frail little guy stare up at each and every airplane in sight and commit himself to reaching those heights as well…. Someday, some way. Today, he is a Captain on the Airbus A-320 for American Airlines and has been flying professionally since 2000. Every member of our family could not be prouder.

I also thank my son-in-law, Knut Butzinger, who offered some very insightful comments after his reading of an early manuscript.

Other much-appreciated support has come from my Mike Festino, Jay Biggenwald, Dee O'Hara, and Alan MacGamwell.

I'd like to acknowledge the talent, dedication, and hard work of the American and international scientists, engineers, technicians, managers, astronauts, cosmonauts, and their families who have contributed to the progress of "Space – Ever Farther, Ever Faster – Now!" and laid the foundation on which we are now building.

An outstanding debt of gratitude is owed to those space professionals who have read versions of this book and provided thoughtful comments and helpful suggestions, which significantly improved its accuracy and readability. They are: (1) Neil Hutchinson, a Lead Flight Director on Skylab and other NASA programs, (2) David Shayler, who has also written extensively about almost all of America's human space programs, women in space, the Hubble Space Telescope, as well as the Soviet space program, (3) Dwight Steven-Boniecki and his wife, Alexandra, who have written extensively about the Skylab Program and recently produced an excellent movie also about Skylab, "Searching for Skylab", (4) Frances French, a good friend, who has excellent knowledge of our steps into space and the English language, (5) Greg Carras, a personal friend with a keen eye and insight, (6) Jerry Carr, the Commander of my mission, Skylab III, who has offered helpful suggestions, (7) Lou Aronica, who has an extensive background in book publishing, published two of my novels and provided a thoughtful comment on this work, (8) Franklin Change-Diaz, who reviewed my discussion of his VASIMR Rocket Engine as well as the remainder of the book and (9) Robert Zubrin, who validated my summary of his input on the

Lunar Gateway. And I also give many thanks to Andrew Chaikin, who read this book several times, and reminded me that I must include much of my own perspective wherever possible, and it naturally fits. That is, and most important, separate this book from a lengthy Wikipedia article by including much of my own experience that I have been fortunate to receive.

Please see Book References in the Appendix for some of the relevant works of Dave Shayler and Dwight Steven-Boniecki.

Throughout the book smart children of all ages appear to inject relevant, human wisdom not yet warped by "superior" opinions from their elders. I put great confidence in their opinions because they are my children: Jannet, John, Jules, and Joe and my wife, Julie.

I acknowledge the highly competent message articulation and content formatting of April Matthews, who, unfortunately, has had to leave our editing team. However, an equally competent professional and cheerful person, Rachel Austen, has stepped in and replaced April. I also greatly appreciate the work of Taylor Edwards in reformatting the book into its current larger and more readable format and additional support.

I greatly value the very thoughtful Forward (below) contributed by Jack Schmitt, also a personal good friend, very competent work colleague, and fellow astronaut. It supplies a valuable background and context for the discussions that follow.

The title "Space – Ever Farther, Ever Faster – Now!" as used here, refers mainly to the manned side of our space program because of its extensive contributions and my familiarity with it. However, it is fully recognized and appreciated that the unmanned side of "Space – Ever Farther, Ever Faster – Now!" is also highly important. It has supplied our primary understanding of the universe in which we live, has many exciting stories to tell, and deserves significantly increased interest and support!

Lastly, but clearly not least, I thank America for the exceptional freedom, opportunity and support it continues to provide.

Ed Gibson

Jack Schmitt, Apollo 17 Lunar Module Pilot on the Moon with Earth above

Foreword

Harrison (Jack) H. Schmitt, PhD

Apollo 17 Lunar Module Pilot and Geologist

Former United States Senator (NM)

Apollo and humankind's further movement into space away from Earth constitutes one of history's most significant milestones. Armstrong's landing at Tranquility Base on the Moon defined a benchmark in human cultural and technological evolution comparable to our species harnessing fire, venturing onto the oceans, creating the American frontier, the British democratic and industrial revolutions, and the Wright Brothers first powered flight in the atmosphere. With the "New Ocean of Space" before us, unlimited new opportunities, known and unknown, loom before our species, both for its advancement and survival.

Astronaut Ed Gibson's extraordinary effort, "Space – Ever Farther, Ever Faster – Now!", recounts the path to this Apollo milestone and America's subsequent adventures and future opportunities in space. The Mercury, Gemini, and Apollo Programs constituted essential factors in ending the Cold War, the mind-boggling degree of the technological and organizational challenges these programs presented, the remarkable advances in science and engineering resulting from success, and what the future in space can mean to humankind. However, the underlying message of this review struck me - taking long-term advantage of reaching a goal is as critical as the goal itself.

Equally important as the enumeration of the lessons of history is Dr. Gibson's tracing of how the nation's leadership gradually lost recognition of why space, intense space, will dominate the advance of freedom and civilization. On the other hand, Gibson points out that some current leaders in government and business appear to have regained a sense of the importance of space access, exploration, utilization, and its essential role in national security.

His most telling statement on this issue is "…The United States must remain first in space…, not just to propel our economy and secure our nation but, above all, because the rules and values of space will be written by those who dare to get there first and the commitment to stay."

Gibson's discussion of the potential menu for future deep space activities is particularly illuminating in "Space – Ever Farther, Ever Faster – Now!". He clarifies that the Moon is on the critical path for exploring Mars, the next great physical objective for humans in space. However, he also notes that NASA must simplify its current approach to a return. Equally illuminating is his discussion of lunar resources in the contexts of reaching, exploring, and eventually settling the Moon and Mars and the potential economic and environmental importance of lunar Helium-3 fusion power for Earth.

Between the spring of 1965 and winter of 1973, Ed Gibson became a good friend as colleagues and active astronauts. His intellect, skill, and good humor are second to none. After our initial training as T-38 jet pilots, from which he graduated second in our Air Force Class of 67A, our astronaut selection Group 4 entered a spaceflight training environment largely biased against having professional scientists and engineers as potential crewmembers. Nonetheless, we persevered in making positive and indispensable differences in the Apollo and Skylab Programs' geopolitical, operational, training, and scientific success.

Clearly, Ed contributed more than anyone rising to the current realization that you send a professional when you need a professional job done!

With this valuable book, Ed Gibson further extends his enviable contributions into the future.

Jack Schmitt

Apollo 17 Lunar Module Pilot

Former US Senator (NM)

Scope

Here we are, you and I, on our backs as we wait for the rocket behind us to explosively light. Each of us feels our heart as it pounds in our ears, senses high-rate data as it streams into our brain and anticipates the future exploration of new worlds up close and personal.

WE Humans are about to Enter Space!

Yes, I felt it as I was about to lift off on Skylab III with my two crewmates, Commander Jerry Carr, and Pilot Bill Pogue. But that was many decades ago. Today we look to either side and sense a "few more crewmates: ten, one hundred, a thousand, a million, a billion, even eight billion more, the whole population of Earth now comes along.

All aboard!

We realize that we no longer travel as single humans but as "WE Humans" that collectively have become "The Human" and we travel uniformly as just **one race**: **The Human Race!**

Finally, we have broken the chains that have bound us to Earth. Finally, We Humans are out of our cradle and can explore and live in the infinite universe around us. This milestone is a big one, one that We have waited 4.5 billion years to reach.

Our excitement as we enter "SPACE – Ever Farther, Ever Faster - NOW!" is not diluted by the number of us on Earth that contribute in one way or another to support this mission — but multiplied manifold!

In America, we first entered space, not so much because of the awareness of the enormous step we had initiated but more because of the practicalities of meeting our international competition resulting from the shock of Sputnik.

Please understand that this is not fiction; it's more exciting than that! It is about what is REAL, everlasting, and what will occupy many of humanity's higher goals and passions far into our future.

It is our call to action!

It is a thrilling story of Human's most exciting adventure written by someone who started at the age of eight drawing pictures of our solar system with crayons on brown paper bags while lying on his living room floor. Someone who, surprisingly, after a very slow start, eventually moved into the thick of it: first Sputnik and America's dynamic response to it, then our multiple landings on the Moon, Skylab, our Space

Shuttle and Space Station developments, and now our upcoming human explorations of Mars and recently discovered exoplanets outside our Solar System. Although American activity is primarily discussed, whenever any human launches from anywhere on Earth, they enter the active "Space – Ever Farther, Ever Faster – Now!"

At each step of the way, there has always been an overwhelming supply of humans who come forward to compete for a limited number of spacecraft seats, then train, fly and perform with competence. Thus, an Astronaut or Cosmonaut might take on many different identities, just as can any person on Earth: nationality, race, gender, ethnic origin, language, beliefs, wealth and body size, type, and capabilities. But regardless of these qualities, there are only *two firm and absolute requirements: the ability to cooperate and an exceptional drive to succeed.*

We in America have also been motivated by our need for international security, technological strength, pride, and drive to explore. However, these ambitions extend into the long-term, and our accomplishments have been short-term, only small steps to date. Yes, WE do have our feet wet, but as WE struggle to get in just over our ankles. WE glance up to see the never-ending ocean of space before us, the unending excitement and opportunities, struggles, and rewards that await us all.

Also, the title "Space – Ever Farther, Ever Faster – Now!" as used here, refers mainly to the human flight side of our space programs because of its extensive contributions and my familiarity with it. However, it is fully recognized and appreciated that all humans who do not fly but make it all possible are and an equally vital part of "Space – Ever Farther, Ever Faster – Now!".

All this space team on the ground has contributed to our MAJOR understanding of the universe we live in, has many exciting stories to tell, and deserves continued and much stronger support.

WE have yet to understand precisely how WE will make up-close and detailed observations of sites light years away to determine if life was or can be supported there. And how can WE ever travel there in practical lengths of time and initially return? WE Humans have made only a very humble start.

It's time. Let's get on with it!

Dedication

To Joe

Joe is our son, who we lost on December 29, 2009. His spirit, drive, and accomplishments were on a par with the best of those who contributed to our advancements in space.

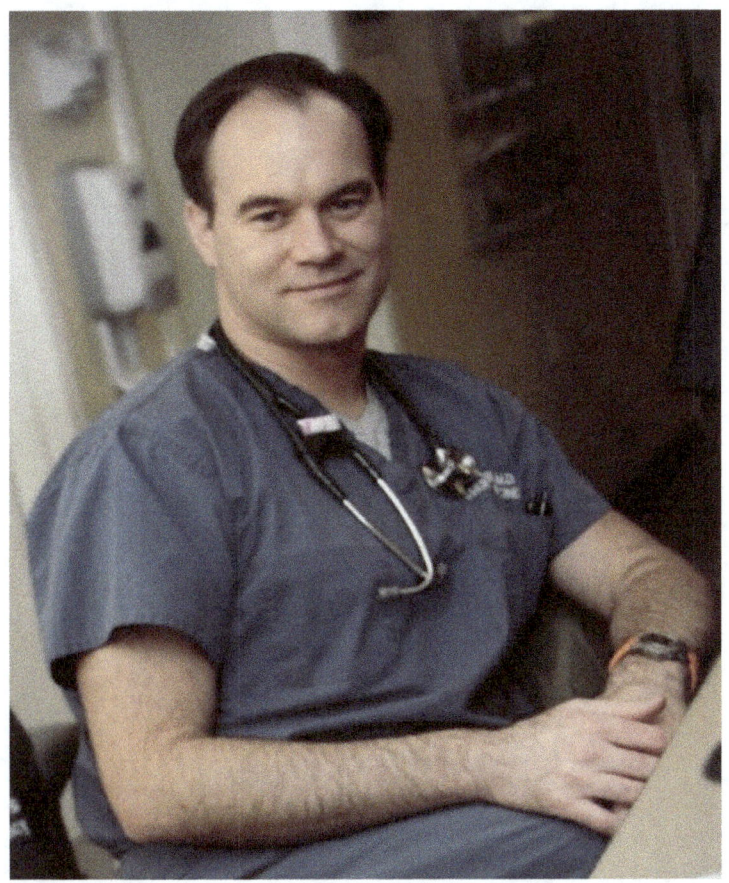

But rather than space, Joe chose the challenge of becoming the best emergency room physician he could be, a goal he achieved far beyond his or anyone's rightful expectations. His eminent professional capabilities, moral strength, and high-level physical conditioning acquired in many pursuits, including triathlons, his relentless drive, warm personality, and strong sense of humor were all wrapped up in one beautiful human that our family greatly loves and misses.

Remember that our love and pride follow you forever regardless of where you are, Joe.

Chapter 1
America's Initial Impulse

WE, the Human, has finally left our cradle and is moving out — fast!

We Americans entered space" when a challenge exploded right over our heads, and we responded!

After Sputnik, our first step outward started with great enthusiasm and accomplishment. Then we became discouraged with our lack of follow-through. Yet today, we still look forward —— hopeful!

> *Do we have sufficient vision, courage, and commitment?*

Then, once again, not a proud moment. Two and a half years after Sputnik on May 1, 1960, came another international embarrassment for America with the Soviets when Gary Powers in one of our U-2 spy planes was shot down over Russia. Nope, not good.

Ed Gibson

We had to get into the race, and we just had to do whatever we could!

On Jan 31, 1961, we quickly got back into the picture when we launched a young, enthusiastic American on a highly successful sub-orbital mission.

Just doing what we could!

America's First Entry into Space!

Then the Soviet Union did it for real when they launched Yuri Gagarin on a successful, "one-orbit, first-human-in-space" mission on April 12, 1961.

Credit: Huntsville Times

Credit: Soviet Image

In the Soviet Union, they celebrated. In America, all we could do was watch from the sidelines and struggle forward at the fastest possible pace.

Few doubts now existed: America was in a heated arms race driven by the urgent need to demonstrate superior technical capabilities to launch rockets that could strike distant enemy targets and defend our homeland. Our damaged egos were terrible enough, but the potential large-scale loss of life would be a much greater tragedy.

> **Can we ever aggressively expand into the world around us…**
>
> *to explore, appreciate and utilize?*

Chapter 2
Another Low for America

On the island of Cuba, just 103 miles south of Florida, a brutal dictator, Fulgencio Batista, repressed freedom for seven years until he was overthrown on New Year's Day 1959 by an even a more brutal dictator, Fidel Castro, who first masqueraded as a liberator but quickly aligned with the Soviets.

Credit: U.S. Government/CIA

The result of a weak U.S. response at the Bay of Pigs - April 17, 1961. Tragically, there was a loss of life for many of the exiles involved.

Space, Ever Farther, Ever Faster – Now!

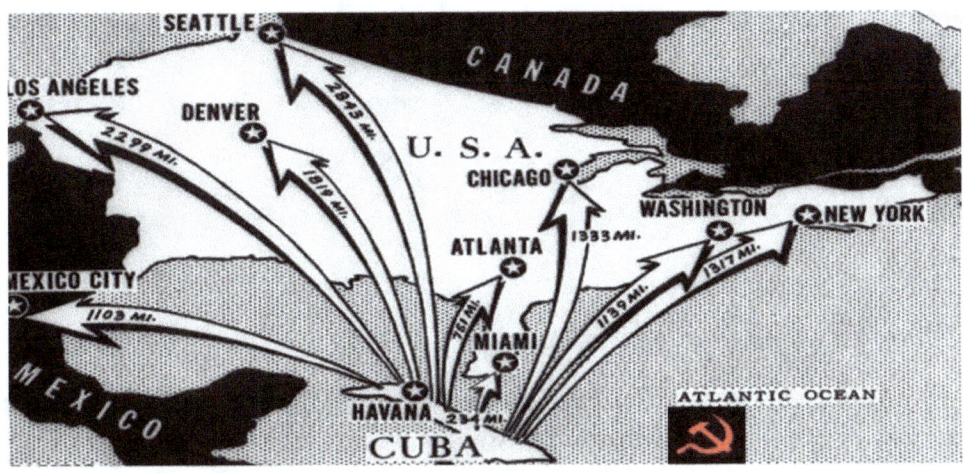

The Cuban Threat

Could many of us get killed by a missile aimed us? Could we lose our lives? Could we lose our Nation?

Credit: Katehorn.com

Fidel Castro

Unfortunately, our new president, JFK, half-heartedly implemented the plan, which was passed down from President Eisenhower, but without the specified American military support. Since then, the Castro regime has remained in power right up to the present day.

Credit: history.com

`*President Kennedy*

After Sputnik, our first steps outward started with great enthusiasm and accomplishment. Then it slowed.

> *Can we adequately respond to the aggressive challenge right before us?*
>
> *Do we have sufficient vision, courage, and commitment?*

Then, once again, not a proud moment.

Two and a half years after Sputnik on May 1, 1960, came another international embarrassment for America with the Soviets when Gary Powers in one of our U-2 spy planes was shot down over Russia. Nope, not good.

We had to get into the race; we had to do whatever we could!

On Jan 31, 1961, we quickly got back into the picture when we launched a young, enthusiastic American on a highly successful sub-orbital mission.

Just doing what we could!

America's First Entry into Space!

Space, Ever Farther, Ever Faster – Now!

Then the Soviet Union did it for real when they launched Yuri Gagarin on a successful, "one-orbit, first-human-in-space" mission on April 12, 1961.

Credit: Huntsville Times

Credit: Soviet Image

In the Soviet Union, they celebrated. In America, all we could do was watch from the sidelines and struggle forward at the fastest possible pace.

Few doubts now existed: America was in a heated arms race driven by the urgent need to demonstrate superior technical capabilities to launch rockets that could strike distant enemy targets and defend our homeland. Our damaged egos were terrible enough, but the potential large-scale loss of life would be a much greater tragedy.

> **Can we ever aggressively expand into the world aro us…**
>
> *to explore, appreciate and utilize?*

Chapter 3
America's Golden Era in Space Begins

The Soviet challenge was immediate and right in our face.

A lesser nation and president would have let this string of embarrassing and dangerous performances continue. But acceptance of continuous failure was not in the nature or grit of the American people and its space workers. On May 5, 1961, 23 days after the flight of Yuri Gagarin, Al Shepard became our first American in space when he made a 15-minute sub-orbital flight. In America, all of us celebrated and took tremendous pride in this great accomplishment of America's first step into space!

The leader of the Soviet Union called it "just a flea hop."

Al Shepard liftoff, extraction from the ocean and receiving medal

However, now with technical confidence based on the excellent work of George Low, NASA's Chief of Manned Spaceflight, and his NASA committee, our young president's reaction became far from half-hearted. He announced a decision he had made even before Al Shepard's flight. On May 25, 1961, addressing a joint session of Congress, President Kennedy boldly proclaimed to the world:

"I believe that this nation should commit itself to achieve the goal, before this decade is out, of landing a man on the Moon and returning him safely to Earth."

President Kennedy delivering his "Moon Speech." Before Congress

"I believe that this nation should commit itself to achieve the goal, before this decade is out, of landing a man on the Moon and returning him safely to Earth."

We stared at one another in America, "We're going to do WHAT?" Also, many in the astronaut corps were glad to see that he added the last six words.

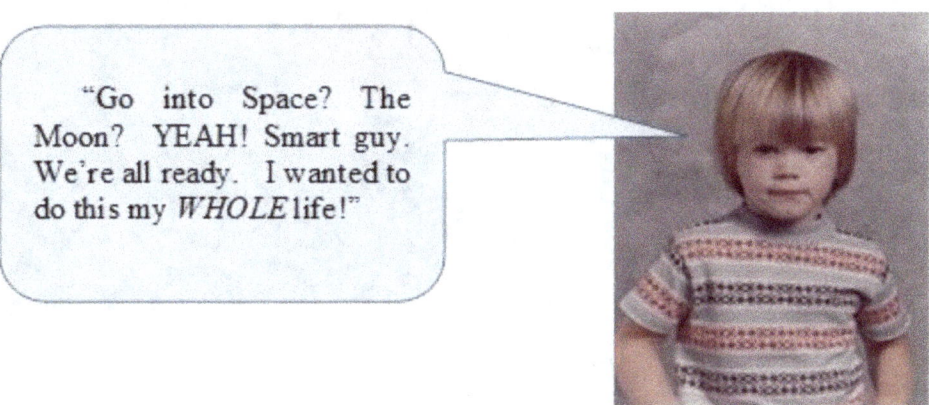

"Go into Space? The Moon? YEAH! Smart guy. We're all ready. I wanted to do this my *WHOLE* life!"

The President's statement opened the eyes of many Americans to a human exploration dream that was about to move into a new arena, into the third dimension, literally. Some science fiction was about to become a reality!

Fires were kindled in the hearts and minds of those who perceived our future and its unlimited possibilities.

Though still in graduate school, I was one so stimulated. A year earlier, I had received a master's degree in engineering with a specialization in Jet Propulsion and Rocketry from the California Institute of Technology. The road before me could not have been more clear or attractive.

Independent of whatever noble ambitions our president expressed as America was poised to Enter Space, the ever-repressive Soviets were at it again. In October 1961, they closed off the last escape route from the East to the West as they walled in the people of East Berlin after only 20% of them had escaped. Apparently, the worker's paradise that the communists created in East Germany was not very attractive to those privileged to live there. This loss of freedom by so many people was viewed as highly regrettable and the Soviet's action as just another in-your-face challenge to the West.

Credit: Smithsonian.com

Russian Challenge at the Berlin Wall

In a speech at Rice University on September 12, 1962, President Kennedy doubled down on his previous pledge. "We choose to go to the Moon in this decade and do the other things, not because they are easy but because they are hard, because that goal will serve to organize and measure the best of our energies and skills because, that challenge is one that we are willing to accept, one we are unwilling to postpone and one which we intend to win!"

All of us celebrated his unconditional strong commitment to taking on "*that which is hard*" for our Nation, especially my fellow Caltech students and myself. This commitment has yielded a string of powerful results as We, the Human, Entered Space. We in America were motivated by the exciting enticement of humanity's first entry into space and our need to win our war of survival against the Soviets.

Wernher von Braun and President Kennedy

Kennedy viewing an Apollo Saturn IB Rocket

President Kennedy really echoed his own growing convictions. Regardless of the evident international and political pressures behind our initial thrust into space, over time, he had also become both intellectually and emotionally fully invested in the heart of our space effort, just as I and many others around me had!

THE FOLLOWING DAY OUR PRESIDENT WAS ASSINATED!

The loss was felt immediately — sharp and painful — by every one of us, our family, our friends, all of us in America. And importantly, over the years, we have come to recognize the significant loss of practical yet visionary leadership of our Nation and especially of our fledgling space effort.

> **With our President's strong leadership gone in a flash,**
>
> *could we ever regain our drive, energy, and momentum?*

Ed Gibson

Chapter 4
What President Kennedy Left Behind

We now faced the difficulty of the loss of President Kennedy's leadership.

Under President Kennedy's leadership, Humanity's view of itself had exploded outward.

Obvious commitments from President Kennedy include:

Political and practical
Initially: The most obvious ones are that at all costs, the United States had to prevent severe international aggression and the threat of a hot war through a show of technical strength and the political will to use it if required. We also had to re-establish American prestige around the world. Lastly, the Administration had to regain political and public support at home.

Credit: govbooktalk.gpo.gov

Noble and Inspirational
Later: I believe that, over time, more altruistic goals also emerged. America found pride and excitement in leading humanity off our planet to the Moon and farther out for the first time ever. We realized that our space effort increased our national strength by spurring significant growth of our technology and economy at home.

Credit: govbooktalk.gpo.gov

Also, there was a growing recognition of the long-range positive impact to our Nation by inspiring enthusiasm for science, technology, engineering, and math (STEM) within our youth. But which of these many factors was it?

It was ALL of them woven together into something called "**Leadership**".

But just what is THAT?

Ed Gibson

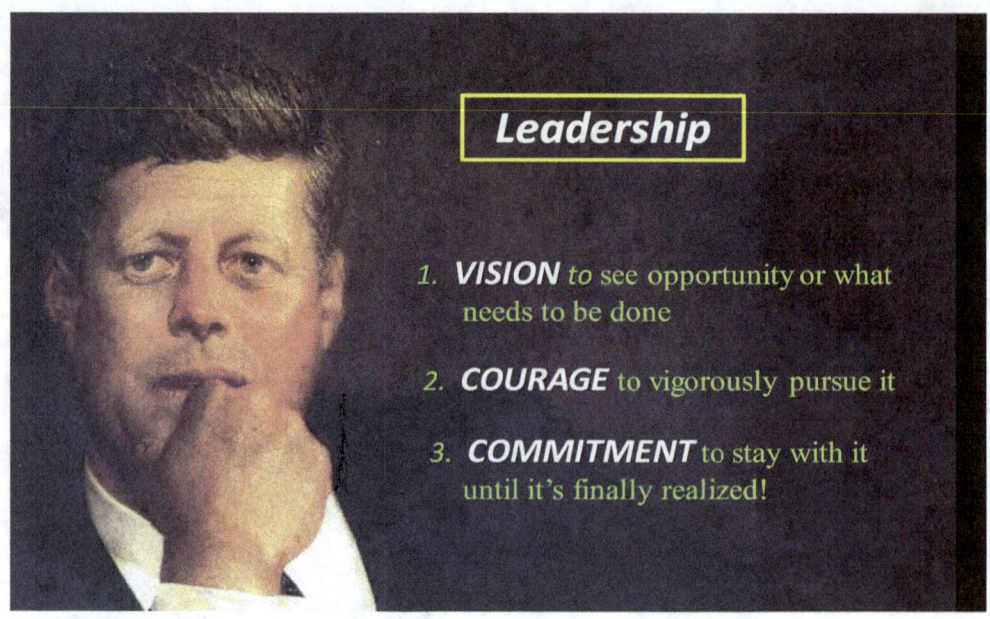

Chapter 5
Mercury – Making It Happen

After World War II, Joseph Stalin, the leader of the Soviet Union (1941-1953), sought to remain competitive in the nuclear arms race with the United States but did not have any bases in the western hemisphere from which to deploy bomber aircraft. Thus, he developed intercontinental ballistic missiles, which eventually led to Sputnik.

Their missiles were real and in our face. The space race between the Soviets and us was on, and that is when we initiated Project Mercury.

Credit: U.S Government/CIA

The Original Seven — American's first astronauts (1959): Scott Carpenter, Gordon Cooper, John Glenn, Gus Grissom, Wally Schirra, Al Shepard, and Deke Slayton

When I graduated from college in 1959, seven of the Americas' best test pilots had just been selected to fly in Project Mercury eight months earlier. These first astronauts indeed led the way for every one of us and the many more who have followed. `

At last, flight in space was taken out of the realm of science fiction and, with the flash of light from a rocket exhaust, had entered the realm of reality.

I still did not have any indication of what my future would hold. However, I obsessively followed each Mercury mission because I sensed that all Americans were on board; NASA had entered space for all of us!

On May 5, 1961, Al Shepard flew the first Mercury mission in his Freedom 7 spacecraft and became America's first astronaut in space. Pride in America immediately took a major step upward.

Virgil (Gus) Grissom flew the second identical mission in his Liberty Bell 7 spacecraft two and half months later. These two shakeout flights were intended to verify Mercury's design and construction.

Both missions were flown on Mercury-Redstone rockets; all the remaining missions were flown on Mercury-Atlas rockets, which enabled them to enter orbit.

On February 20, 1961, John Glenn flew the third mission in his Freedom 7 spacecraft, which lasted for three orbits of Earth. It was a highly celebrated mission—even more than Al Shepard's first flight! Then Scott Carpenter flew in his Aurora 7 spacecraft and, like John Glenn, covered three orbits. Because of technical and human errors, he landed 250 miles beyond his intended landing site.

His mission lasted for nearly four hours, but the final two missions were even longer: Wally Schirra flew his Sigma 7 spacecraft for over 9 hours, followed by Gordon Cooper who flew the final Mercury mission that lasted over a day. Seven astronauts were selected, but only six flew.

Whatever happened to Deke Slayton?

Deke Slayton

Unfortunately, in a routine physical exam while in training, the doctors detected an "off-nominal" electrocardiogram reading. They continued to observe and debate. Finally, Deke was grounded and not permitted to fly until much later in the Apollo-Soyuz Program.

Until he was again cleared for flight, Deke took over all NASA's human flight operations for many years. His strong instinctive leadership was needed at that time, and American Space was the beneficiary! Once he did fly, there were smiles and cheers all around

Also, of lasting value from the Mercury Program, six of the Original Seven Astronauts had the wisdom to initiate, the Astronaut Scholarship Foundation (ASF). The seventh member of the group, Gus Grissom, who was deceased by that time, was positively represented by his wife Betty. Two other friends and supporters, William Douglas, M.D. and Henri Landwirth, also joined in.

Over the past 40 years, the ASF has provided over $4 million in financial support for over 400 outstanding Science, Technology, Engineering and Math (STEM) college students who attended 45 different universities in total. Since its inception in

1984, it has grown more extensive and effective every year. Each Astronaut Scholar has received up to $15,000 in support. Over my career, I have had the honor of presenting a few of these awards to highly deserving recipients.

Chapter 6
Gemini – A Steppingstone to Greater Spaceflight Capabilities

Next was a bridge t0 the Apollo Moon Landing, Project Gemini, which flew ten human-crewed missions. Two more astronaut groups were added to the first group to man Gemini and the following programs: Astronaut Group 2 (September 1962 – 9 members) and Group 3 (October 1963 – 14 members).

The Program, Gemini is Latin for "twins", and initiated in 1091 and concluded in 1966.

Capabilities necessary for Moon landings were developed and demonstrated.' These included the design, implementation, and demonstration of the new hardware and utilization techniques required to go to and return from the Moon.

Thee **OFFICIAL** Gemini Program Mission Objectives

"OK Guys, <u>*GET THIS STRAIGHT*</u>!

First, we went into space many times, then stayed up for two weeks. Now, two of us will always go up together and a couple of times one of us goes outside. Another time two of us join up with two more of our buddies and dock. Child's play.

You got it? Good. Let's get on with it and make us ready to go to the Moon!"

The **REAL** Gemini Program Mission Objectives

1. To demonstrate endurance of humans and equipment in spaceflight for extended periods, at least eight days required for a Moon landing, to a maximum of two weeks
2. To effect rendezvous and docking with another vehicle, and to maneuver the combined spacecraft using the propulsion system of the target vehicle
3. To demonstrate Extra-Vehicular Activity (EVA), or space-"walks" outside the protection of the spacecraft, and to evaluate the astronauts' ability to perform tasks there
4. To perfect techniques of atmospheric reentry and touchdown at a pre-selected location on land

In landing on the Moon before the decade was out, United States would catch up and overcome the lead that the Soviet Union had earned in our race with them. Thus, in the Gemini Program, we needed to demonstrate a mission endurance of up to 12 days, the time it took to go to the Moon, perform work there, and then return. We also needed to develop and demonstrate methods of performing work during extravehicular (EVA) on the Moon as well as the orbital maneuvers necessary to rendezvous and dock with another spacecraft.

All Gemini flights were launched on a Gemini-Titan II, which was a modified Ballistic Missile, from Launch Complex 19 (LC-19) at Cape Kennedy in Florida. ALSO. Gemini was the first program to use the first completed Mission Control Center at the Houston Manned Spacecraft Center for flight control.

The astronaut corps that supported Project Gemini included the "Mercury Seven", "The New Nine" and "The Fourteen". Unfortunately, during the program, three astronauts died in two air crashes during training, including the prime crew for Gemini 9. The backup crew then flew the mission.

Space, Ever Farther, Ever Faster – Now!

Credit: Wikipedia

Inside the Gemini Cabin

Credit: Wikipedia

*Ed White on first American
Space Walk on Gemini 4
June 3, 1965*

Credit: Wikipedia

*Dick Gordon and Pete Conrad
Gemini 12
There was none better*

With the success of the 12 spacewalks, NASA felt ready to execute the Apollo Program and land on the Moon.

Scientists, pilots, and other Americans are pulled into space to explore, learn, and utilize.

On Becoming an Astronaut

Before the end of Project Gemini, NASA announced that they intended to select the fourth group of astronauts. In part, because of the more scientific nature of upcoming lunar missions than previous manned space activities, the selection criteria considerably emphasized scientific knowledge more than test pilot experience.

At that time, I was just out of graduate school at Caltech for less than a year. Over breakfast one morning, Julie read an article to me from the Los Angeles Times that stated NASA was looking for scientists who wanted to fly high-performance aircraft and fly in space. I thought she was making it up because she knew exactly what I would like to do. But her wording got too precise, too bureaucratic, which told me she was reading a genuine article.

Immediately, we started to think about it in detail, for it would require some analysis, a drastic career change, a different skill set with considerable commitment. We carefully analyzed all aspects of it for a good length of time; then, I applied at 8 AM that morning.

I needed a new significant challenge, and the Space Program provided it. Over the past 12 years of schooling and work, I had been in the continuous all-out, full-on work mode. After flunking 1st and 4th grades, I had started to turn it around. Granted, I was in hospitals and wheelchairs when I was sick with osteomyelitis in my right shin (ages 2, 4, and 8), an infection in the bone that makes it locally soft to the point that it becomes structurally inadequate and causes a high fever.

So, of course, I was afraid that I would never pass the physical because of my history of osteomyelitis. It was clear that NASA did not need any one-legged astronauts. Fortunately, in 1928 Sir Alexander Fleming, a Scottish researcher, discovered penicillin that by 1938 was an accepted medicine and its use well understood. For me, it worked!

I probably used my sickness as an excuse to justify my poor grades. But my Dad wouldn't accept excuses. Every six weeks, when I brought home a report card filled with D's and E's, my Dad and I had an "encouragement" session, after which I could not sit down and had no doubt what my Dad was demanding.

This was my first brush with tradeoffs, which I later used in my engineering studies. It was also when I started to develop my analytical skills. After every encouragement session with my Dad, I pondered, "If you continue to screw around, then every six weeks, you will have another painful encouragement session. And each session will seem more painful than the previous one."

"But, if you work harder, then every six weeks there will be no pain, and everyone will be happy."

Hmmm, much to think about.

For a few more years, my answer was, "No, I think I'll screw around for next six weeks, and then take my medicine for a few days."

My Dad, who was very determined and who I loved very much, made it clear that he would never give up on me. The message sunk in that I should not give up on myself either. I started working harder and made a fantastic discovery. "If you work harder, your performance improves."

"What? Really! Who knew?!"

By the start of high school, my performance skyrocketed all the way up to mediocre! By about the middle of my junior, I was receiving mostly A's and B's and realized that college was my last opportunity if I didn't want to do those summer jobs for the rest of my life. So, I moved the importance of my studies right up there next to football and track.

When I applied for college, Cornell didn't want me to even walk by their campus (a slight exaggeration). A few other colleges would take my money. But, by far, the University of Rochester was my best opportunity. After my Dad and I visited the Director of Admissions, all my Dad could say was, "It's a long shot but worth a try." Six weeks later, I received a letter of acceptance from the U of R. It made my day—and my life!

After six weeks at the U of R, a friend told me that he saw my name on the Dean's List. "What? I never even met this Dean guy! What did I ever do to him? Why am I on his list?

About ten years later, when the U of R chose to award me, an honorary Doctorate based on my records in space and academics, I felt grateful that they decided to bet on me again.

After the University of Rochester, I was awarded a National Science Foundation Fellowship that covered all my expenses over four years of graduate school. But what school should I choose?

Since I was interested in jet propulsion, rocket flight, engineering, and physics, I chose the California Institute of Technology or Caltech. Academically, it had the courses that I desired, and its quality was one of the best in the nation. Five years later, I received a Ph.D.

Once back in the ordinary world after graduate school, I got bored very quickly. But I learned that if I made it to the final round of an upcoming Scientist-Astronaut selection, I would get a shakeout ride in a T-38. That alone made it worth the effort!

If your gut and mind tell you that you really want it, and it is good for you and those close to you, <u>go for it!</u>

For nine long months, NASA labored over the birth of the first group of Scientist-Astronauts. In June 1965, after all that time, all that paperwork, all those examinations and interviews, and all those excellent candidates in the approximately 1,100 applicants, NASA could find, I concluded, only six semi-healthy scientists. It turned out that I was in the right place at the right time and just healthy enough.

NASA concluded that since I had not had a reoccurrence of osteomyelitis for 21 years, the bacteria in the bone was dead. Unfortunately, one other candidate with the same background but a more recent infection was not accepted.

Before acceptance, NASA brought 16 of the top candidates down to Houston for interviews, physicals, a wring out in a T-38, and ride on a centrifuge. The ride on the centrifuge simulated the G-loading an astronaut would experience if an abort was made during liftoff on a Saturn booster. This G-loading versus time was a bell-shaped curve with a peak acceleration of 10.5 G, which means that at the peak of the 10.5G on my 160-pound body, I would weigh 1,680 pounds!

(1965 – Six Members) Front – Curt Michel, Jack Schmitt, and Joe Kerwin Rear – Owen Garriott and Ed Gibson – One other candidate, Duane Graveline, left the Program before this picture was taken.

Blissfully, in got strapped in. Never having experienced 10.5 G before, I was in for a surprise. As the "ride" progressed, I could have sworn that my ears were finally getting acquainted with one another behind my head, and my cheeks were not far behind. At the end of the "ride," I stood up straight, as well as I could stand up straight, and walked confidently like I had just gotten off the bunny ride at the amusement park.

"How did it go, Mr. Gibson?"

"Ahh, no problem!" Smile, "Yeah, yeah, no problem! No problem!" Smile. Smile.

In truth, my chest felt like a steam roller had run over me. I was able to suck in just enough air to force out a few words at a time and stand upright by starring at the vertical edge of a wall 10 feet in front of me.

I felt pain for over a week, but I wasn't going to say anything that gave the slightest bit of satisfaction to that high-G, torture community.

When I arrived in Houston before being shipped off to flight school, I found myself standing next to many of my idols up to that time.

Clearly, some of the existing test pilots in the program viewed us as imposters. We suspected that some of the current astronauts approved of our assignment to flight school where any one of three good things could have happened: we could have quit, flunked out, or killed ourselves, and they would never have to deal with us encroaching on their turf again.

With this cheery thought in mind, we headed off to flight school—time to get to work.

Driven by International competition, Humans head to the Moon.

Chapter 7
Apollo – The Pinnacle

Well before the completion of Gemini, it was recognized that Apollo required not only more astronauts, new boosters, and spacecraft of significantly greater capabilities but new supporting facilities with significantly upgraded capabilities as well. The NASA team got to work. Sometimes, stronger adjectives will not do the job in describing these new Apollo capabilities. Often just plain old numbers of greater and greater magnitude must be called upon. The following few pages illustrate this reality.

Saturn Vehicles

A new class of vehicles was developed for the Apollo Program by Wernher von Braun's team at the Marshall Space Flight Center: the Saturn 1 B (224 feet tall with a liftoff thrust of 1,600,000 pounds) and its big brother, the Saturn V (363 feet tall with a liftoff thrust of 7,610,000 pounds). The Saturn 1B was used for early Apollo Earth orbital flights, three manned missions to the Skylab space station, and the Apollo-Soyuz Earth-orbital mission. The Saturn V was used to launch all flights to the Moon and the Skylab Space Station itself into Earth orbit.

The S IVB was the second stage of the Saturn 1B and identical to the third stage of the Saturn V, as were the spacecraft above each S IVB. The Saturn 1B's spacecraft and upper stage were linked to the same connections on the launch gantry that its big brother used. It was boosted up on a "milking stool" (below right).

The big guy—Saturn V   *Little brother —Saturn 1B*

Wernher von Braun effectively led NASA's development of our large rockets. He first worked in Germany on the V-2 during WWII and then at NASA. His technical support to Hitler made him highly controversial once he came to the United States. After all, he had used prisoners to help build the V-2.

All indications were that although he did a credible job building the V-2 during the war, he, and his team of rocket engineers that he brought with him, were not in any way faithful Nazis or sympathizers. My experience with him in the late '60s and early '70s revealed that he was not only a visionary and a competent, strong leader, but he was even stronger because he was a leader with explicit consideration and liking for ALL others—a man who was easy to like and respect. The members of his team and our team felt the same.

Apollo Spacecraft - Command and Service Modules (CSM)

The Command Module (CM) was a three-person spacecraft designed to house crews during the launch into Earth orbit or to the Moon and return them to splashdown on an ocean somewhere on Earth. It's conical or "gumdrop" shape was covered with heat rejection materials, especially the ablative heat shield on the broad bottom surface that shielded the crew from the heat of re-entry at lunar return speeds of 25,000 miles per hour.

Once the CM had slowed sufficiently within the near-earth atmosphere, parachutes were used to slow it further to acceptable splash-down speeds. The CM was mounted on a cylindrical Service Module (SM) that contained an extensive service propulsion system (SPS) and fuel cells that generated electrical power (Please see the picture below.). Both systems used liquid Oxygen and liquid Hydrogen, and the SM also provided Oxygen for the crews to breathe and thermal control for the CM.

The SM had four sets of reaction control jets that supplied both attitude control and translation maneuvers too small for the use of the SPS. It also had a high-gain S-band antenna for long-range communications on lunar missions. The SM was used to put the CM at the right attitude and speed to perform re-entry before the CM separated. Over the years, it did not change from its original shape.

Apollo Command (CM) and Service Module (SM) -The segment, the CM, houses the upper crew.

Apollo Spacecraft - Lunar Module (LM)

The Lunar Module (LM) was propelled into lunar orbit by the three stages of the Saturn V and the SM SPS. Once in lunar orbit, two crewmen entered it, separated from the CSM, and landed on the Moon using the LM descent engine. When lunar surface activities concluded, the LM ascent stage launched the crew from the now inert LM's descent stage.

When it came close to Earth, the CM separated from the SM, re-entered, and splashed down on the ocean. Because the LM did not fly through the Earth's atmosphere or return to Earth with a crew, it was designed without aerodynamic considerations. Also, because of weight restrictions imposed by propulsion system capabilities, it was designed to be as light as possible. Before launch, each crew or technician around it clearly understood that one careless move could lead to a puncture of its thin walls. Visually, it reminded me of tinfoil.

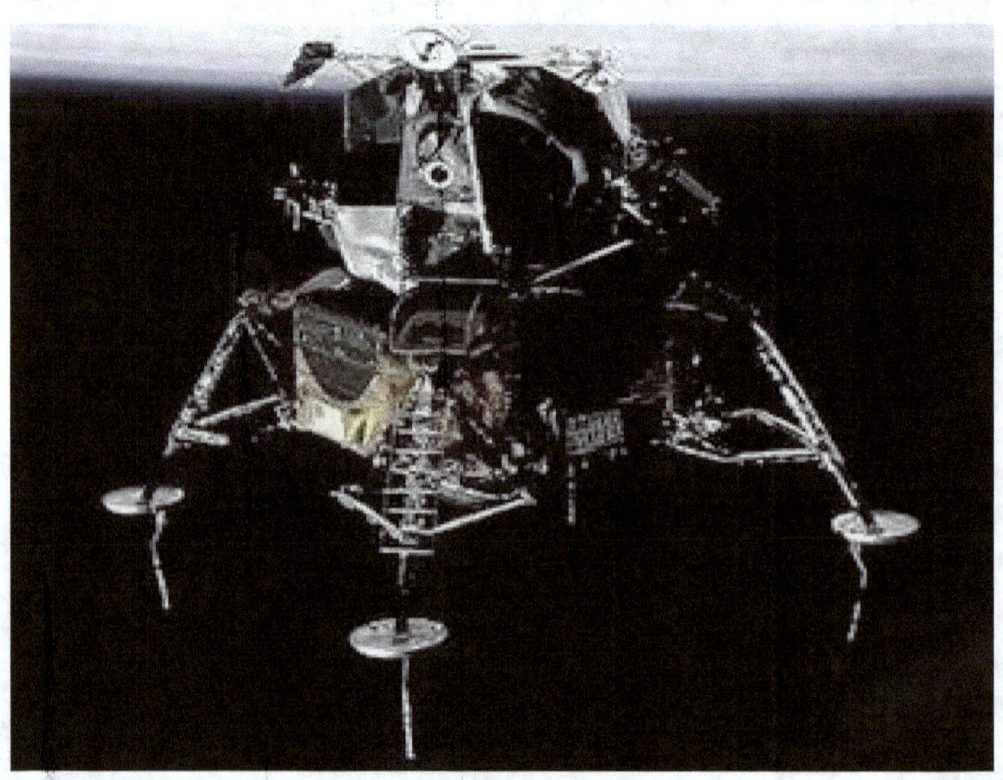

Lunar Module Ascent and Descent stages - Descent stage landing legs extended in an Apollo 9 test.

The Command and Service Modules (CSM) and Lunar Modules in their docked configurations are show on the preceding page. Previously, the LM extra has extracted from the Saturn S IV B stage by the CSM.

Supporting Facilities

The Manned Spacecraft Center (MSC) was constructed to house new and expanded ground and simulation functions and provide adequate program management facilities.

The Center's name was later changed to the Johnson Space Center after LBJ's death. LBJ had made sure that our space effort was properly funded after JFK's assassination. Because of the much larger sizes of the Apollo Saturn Launch Vehicles, their facilities in Florida expanded commensurately.

The new Launch Operations Center constructed on Merritt Island included Launch Complexes (LC) 34 and LC 37 and the largest LC 39, a state-of-the-art Operations and Checkout Building, and a mammoth Vertical Assembly Building that has an internal volume of 130 million cubic feet.

Vertical Assembly Building (VAB) and a Saturn V on the Mobile Transporter (crawler) leaving its birthplace on its way to LC-39A

The Tragedy of Apollo 1

Apollo was ready for its first flight in the nearly ten years since Sputnik and accelerating from an almost dead stop.

*Apollo 1 Crew:
Sr. Pilot Ed White, Commander Gus Grissom, and Pilot Roger Chaffee*

On January 27, 1967, the crew and total system was ready, Suited, the Apollo 1 crew of Gus Grissom, Ed White, and Roger Chaffee were set to test their spacecraft on the launch pad and simulate the countdown they planned to perform on their actual launch date of February 21, 1967.

Apollo 1 on its launch pad

During the test, a fire broke out in the cabin and spread quickly because of the high-pressure 100% Oxygen inside. Tragically, the crew could not quickly open the hatch and escape. These three very good men of Apollo 1 were burned and asphyxiated!

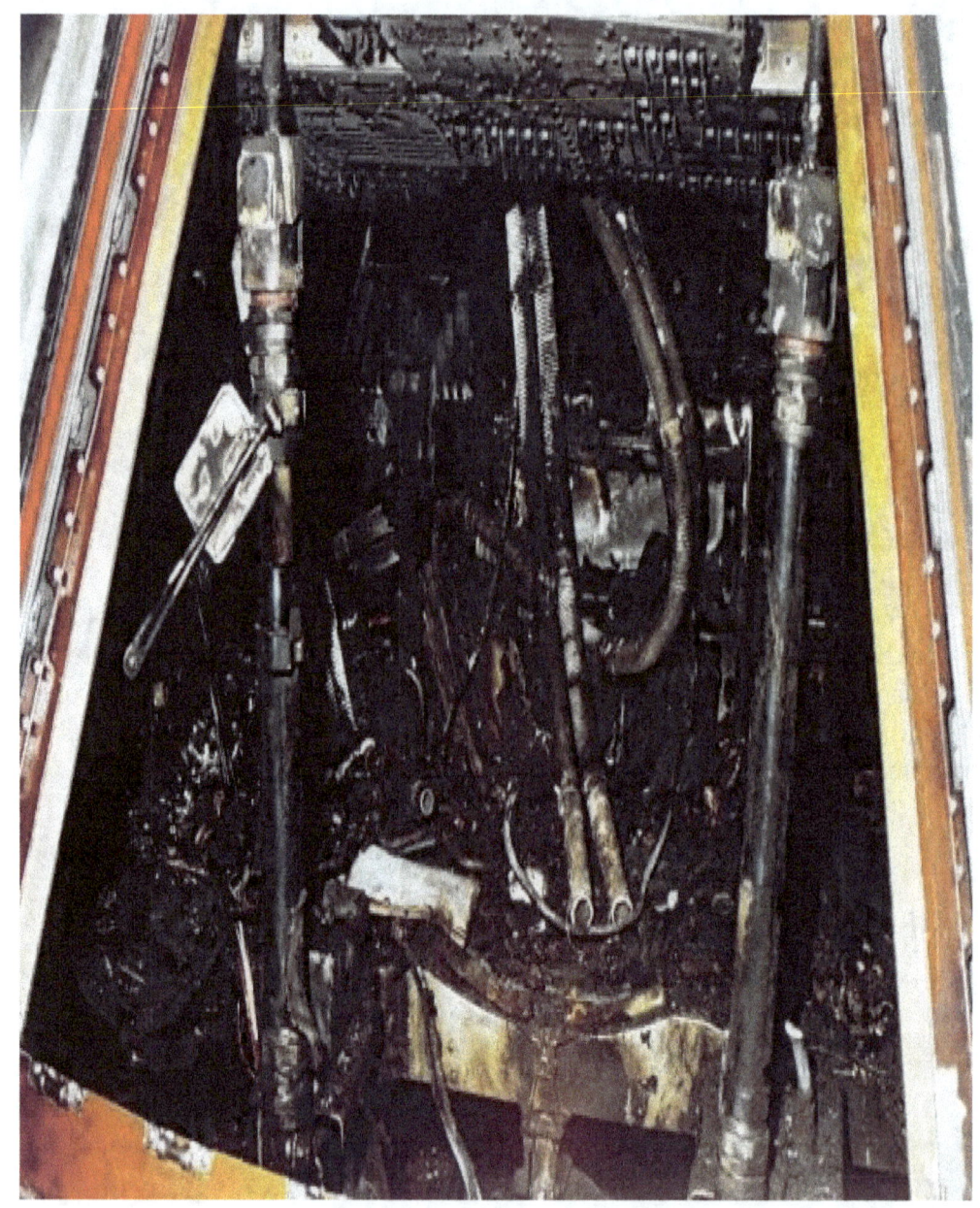

Inside of CM Cabin after crew was extracted.

The Program was brought to a standstill, literally, before it ever got off the ground. As expected, because of its major impact on the program and the gruesome nature of the accident, some pushed forward and demanded, "Is THIS what we're spending all our money on? "Then Stop! Quit! Now!"

But when President Kennedy committed us to reach the Moon before the decade's end, the NASA team performed with determination, growing expertise, and the joy of accomplishment. Some labeled it as "go fever." However, things had changed in the flash of fire, and America was put to one of its greatest tests. Clearly, the Apollo Program and America's commitment were now faced with a considerable challenge.

In a prophetic picture where the crew posed over a model of the CM and prayed to a higher power to fix all the problems that remained in their spacecraft eight months before the scheduled liftoff.

Gus once said, "The conquest of space is worth the risk of life." Fortunately, almost all people in America agreed.

Ed White, Gus Grissom and Roger Chaffee
The crew understood the risk they faced.

George Low

Immediately a critical change was made. George Low (left) took over the Apollo Program. He was not only an excellent engineer and manager, but he was also an exceptional leader. He restored calm, got people working together, and brought a strong focus on what had to be done.

He originated the Change Control Board that promoted rational problem solving by bringing together all relevant people, interests, and perspectives. Their meetings were frequent, long, and detailed.

Also, George held an eye-to-eye meeting with each crew to review hardware status right before their liftoff, a practice he maintained all the way through Skylab. When he briefed us on the status of our spacecraft on Skylab III, his detail was precise. When he said it was ready to go, we were also immediately fully ready to go.

George was a former refugee from Austria whose Jewish family had fled Nazi Germany in 1938. He was a very competent, admirable, and exceptional man in every respect. There were no flashes of anger, a petulant or arrogant attitude, or forcing of others to take a backseat to his ego. He treated everyone with respect, an open mind, and a friendly demeanor. But it was clear that he was always focused on results.

NASA was fortunate to have George Low there in the right place at the right time. Many current and future managers chose to emulate him; unfortunately, a few did not.

Secondly, a review board was convened and led by Frank Borman, a Gemini astronaut who focused on speed, excellence, and meaningful results.

It concluded that deficiencies existed in the Command Module design, workmanship, and quality control. Also, the use of 100% Oxygen, flammable materials in the cabin and suit materials, and a hatch that could not be rapidly opened outward were major contributors to the deaths. With a firm resolve, the program is set to work to make the required fixes and quickly get Apollo back on the right course.

Chapter 8
Apollo 7-13
Apollo 7 – First Manned Flight of Apollo

Apollo 7 Crew - Command Module (CM) Pilot Don Eisele, Commander (CDR) Wally Schirra and Module (LM) Pilot Walt Cunningham

Mission

On Oct. 11, 1968, Apollo 7 lifted off from Complex 34. They entered an elliptical orbit of 140 by 183 miles. The objectives of this first flight were to demonstrate the performance of the CSM, mission support facilities, rendezvous capability, and live TV broadcasts from space. Wally summed it up. "She is riding like a dream."

After one-and-a-half orbits, they separated the CSM from the S-IVB to set up a simulated rendezvous and docking like would be required on future Moon landings. This was accomplished the next day when the CSM and the S-IVB were 80 miles apart. Using two burns of their Service Propulsion System, the crew maneuvered their CSM to within 70 feet of the now lifeless S-IVB.

Despite the nominal functioning of the spacecraft systems, about a half day into the mission, Wally then the other two crewmembers realized that they were coming down with bad colds. They did not get adequate sleep before the flight and the colds moved in quickly. Without gravity these colds were exceptionally uncomfortable because their nasal passages filled with mucus but did not drain as they would in gravity. A few medications they had onboard helped a little, but they had no choice but to press on.

The crew made the first live TV broadcast of American crewman from space. Although the pictures were of low quality, they proved to be of interest to the public and students. They followed it up with six more broadcasts throughout the rest of the mission. These broadcasts led the way for the hundreds that would follow of higher quality and interest on almost all future spaceflights.

They performed eight firings of their Service Module (SM) engine. This engine was crucial for future flights to the moon. At crucial times during a lunar voyage, the engine simply had to work, or they would not get home. On Apollo 7, all 8 firing attempts were nearly perfect. On their first firing of the SM engine, the crew felt an unexpected sharp jolt and then were pressed in their seats for over a minute as the engine continued to fire. The sharp engine ignition was unexpected but proved to be just the nominal SM engine mode of operation.

When the crew splashed down in the Atlantic Ocean southeast of Bermuda, the Apollo team realized that they had qualified the CSM and were now ready to use it on upcoming missions to the Moon. Enthusiasm was high as NASA shifted focus to Apollo 8.

As We Entered Space, we had come roaring back, made that long stride outward, and beat the Russians to the Moon. We were now ready to fulfill America's commitment stated by President Kennedy seven years earlier to land humans ON the Moon and return safely to Earth. About that time, the mood of the Nation was somber with bad news coming from North Korea, Vietnam, Czechoslovakia, and the murders of Bobby Kennedy and Rev. Martin Luther King.

It had been less than two years since the tragic human losses in the Apollo 1 fire. All of us in America were ready for something uplifting that would bind our Nation together. Apollos 8 through 11 supplied it.

Apollo 8

Human's firsthand view of its cradle from the near vicinity of our Moon

Apollo 8 Crew: Commander Frank Borman, LM Pilot Bill Anders, and CM Pilot Jim Lovell

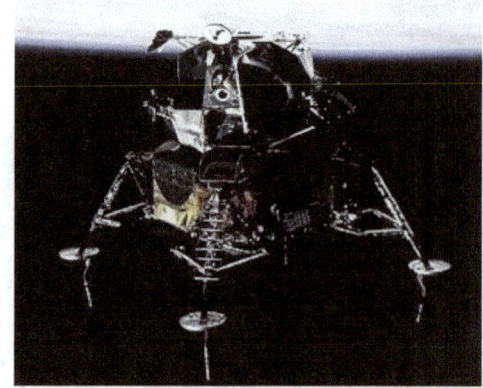

Apollo 9 LM seen from the CSM

Two crewed missions followed quickly to test the total system, including the LM.

Apollo 9

The mission was launched on March 3, 1969, with Commander Jim McDivitt, CM Pilot Dave Scott, and LM Pilot Rusty Schweickart. They remained in Earth orbit and demonstrated rendezvous, docking, LM operations, extravehicular activity (EVA), which is usually just called "spacewalking."

Apollo 9 Crew: Commander Jim McDivitt, CM Pilot Dave Scott, and LM Pilot Rusty Schweickart

Apollo 10 - Lunar Shakedown Mission

Apollo 10, launched on May 18, 1969, with Commander Tom Stafford, CM Pilot John Young, and LM Pilot Gene Cernan on board, was a full rehearsal for a lunar landing except that Tom and Gene took the LM down to only within 50,000 feet of the lunar surface without attempting a landing while John Young remained orbiting overhead in the CSM. The stage was now set for the first humans landing on the Moon!

*Apollo 10 Crew:
LM Pilot Gene Cernan, CDR Tom Stafford and CM Pilot John Young*

Space, Ever Farther, Ever Faster – Now!

Human takes the first big step onto to a celestial body other than Earth.

Apollo 11

Apollo 11 launched on July 16, 1969, and experienced only minor issues up to initiating the burn in their LM to start down to make their lunar landing.

Apollo 11 Commander Neil Armstrong, CM Pilot Michael Collins, and LM Pilot Buzz Aldrin

Then Neil and Buzz saw they were headed to land miles beyond the intended area; computer alarms distracted their attention, and when close to the surface, they were headed toward a boulder field. Also, a low fuel warning indicator came on.

True to form, Neil remained composed, took semi-automatic control, flew over the boulders to a better location, and landed with only 25 seconds of fuel remaining, which turned out to be less than on all following lunar landing missions.

On July 20, 1969, at 10:39 PM EDT five hours ahead of schedule, Neil opened the LM hatch, descended the ladder, and stepped onto the Moon.

*"That's one small step for man,
one giant leap for mankind."*

Human's first step off Earth!

We Americans felt great joy and pride then, just as we still do today. But even more important, the rest of the world understood "We" to mean ALL CITIZENS OF THE WORLD and landing on the Moon was as accomplishment BY ALL HUMANITY.

It is estimated that 20% of the world's population watched as We Humans walked on an astronomical body other than Earth for the first time ever.

For one moment in our history, our world was united!

WE DID IT!

Many people wished that President Kennedy was still alive to see his challenge to America was met.

Sitting 15 feet to my left in the Mission Control Viewing Room were a few NASA managers, including Wernher von Braun. Everyone focused intently on the progress of the landing. When Neil stepped on the moon, I glanced over at von Braun and saw tears running down his cheeks. Undoubtedly, much dynamic history was running through his mind about how this point in human history had been reached.

Space, Ever Farther, Ever Faster – Now!

 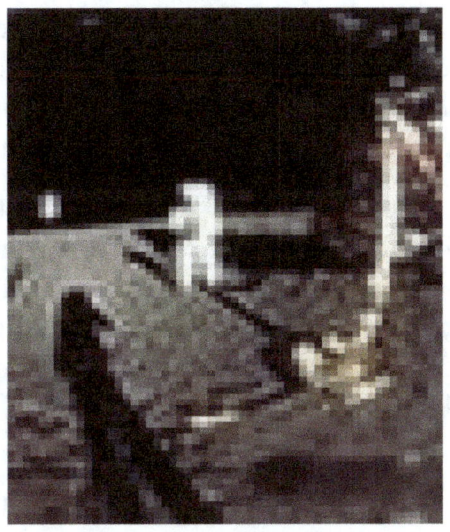

Buzz on the Moon

This picture of Buzz was one of many taken by Neil throughout their time on the lunar surface.

Neil on the Moon

It is a rarity to find a crisp picture of Neil on the lunar surface, even a tiny reflection of him in Buzz's helmet.

Even though they spent only 21 hours 36 minutes on the Moon's surface and performed only one very short lunar walk (Neil - 2 ½ hours, Buzz - 1 ½ hours), they did setup the Apollo Lunar Surface Experiment Surface Package (ALSEP), perform other lunar experiments, and, of course, be the first humans to walk anywhere in our universe other than on Earth.

Neil inside the LM *(picture smaller to accommodate caption)*

Neil's piloting skills, coolness under pressure, intelligence, open friendliness, and modesty made him the perfect American for the responsibilities he carried. I found him always open to meaningful discussions or lending a hand where he thought he could help. NASA and the world need more Neil Armstrongs!

When we think of Apollo 11, we think of the First Man on the Moon. For sure, Neil and his crew were at the tip of the spear and deserved that recognition. But we should never forget that the spear behind the tip has made it all possible, as is true in every one of NASA's human space flights. The whole NASA team should take great pride in what they accomplished and be recognized by the rest of humanity for making it happen.

While still in flight, Mike Collins stated, "We have always had confidence that this equipment will work properly. All this is only possible through the blood, sweat, and tears of many people. All you see is the three of us, but beneath the surface are thousands and thousands of others, and to all of those, I would like to say thank you very much." Neil and Buzz also added their individual words of recognition and appreciation.

Outside of those directly involved or supporting the mission, sentiments expressed were less enthusiastic and usually short-lived. Despite the enormity of this accomplishment for humankind, most Americans eventually whisked their hands together and concluded, "Since we have now been there and done that, it's time to take our eyes off the horizon and return them to the real world right down there at our shoes, which must demand our immediate and constant attention."

Clearly, most Americans consider lunar geology, the historical growth of our solar system, and humanity's evolution as far too abstract to compete with today's immediate earthly demands. Fortunately, a small fraction of people was able to understand the abstract while still meeting their demands of the day.

Interestingly, our fierce competitor in our race to the moon, the Soviet Union, once coming in second, abruptly changed their competitive cry to "landing humans on the Moon is dangerous and unnecessary. We are concentrating on creating large satellite systems." No mention was made of their attempts to send humans to the Moon halted by technical difficulties.

The American flag in the picture (below) was erected by the crew in the spirit of patriotism. Unfortunately, their rocket's exhaust blew this flag over when the crew lifted off. Subsequent missions placed their flags farther away.

Space, Ever Farther, Ever Faster – Now!

Neil at storage compartments in descent stage of Lunar Module

"HERE MEN FROM THE PLANET EARTH FIRST SET FOOT ON THE MOON

JULY 1969, A. D.

WE CAME IN PEACE FOR ALL MANKIND."

Apollo plaque on one landing leg of the Descent Lunar Module left on the Moon

Ed Gibson

AMERICAN ENTHUSIASM RAN HIGH!

American enthusiastic patriotism on full display in this Apollo 11 victory parade like many others to follow.

Apollo 12 Spaceflight with Skill, Accomplishment, and Joy

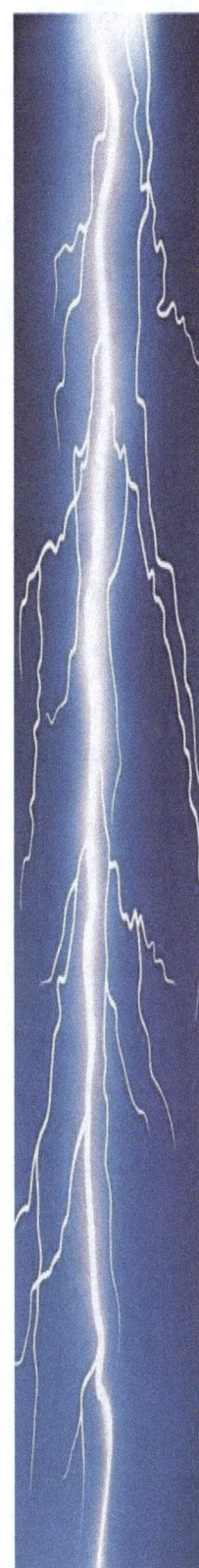

On November 14, 1969, Apollo 12 lifted off into a rainstorm. Thirty-seven seconds later, the vehicle and its hot ionized exhaust plume became the world's tallest lightning rod when a bolt of electricity surged through them to the ground. Protective circuits took all three fuel cells offline and much of the CM instrumentation. A second strike at 52-seconds knocked out the attitude indicators in the cockpit, and most of the instrumentation readings sent to the ground were garbled. Fortunately, the vehicle's guidance was using the Saturn V's Instrument Unit data, which remained on course.

Inside the cockpit, the loss of power lit up almost every Caution and Warning indicator and knocked out nearly all displays. The crew knew they were going somewhere really fast, they just didn't know where!

The memories and actions of two individuals saved the day. John Aaron, the engineer in Mission Control in charge of the electrical, environmental, and consumables systems, knew how to switch the instrumentation over to its backup power supply and suggested that the Flight Director request, "SCE to Aux." Very few people understood where the SCE switch was or what it did. In the cockpit, Pete and Dick were also ignorant.

Fortunately, Al did remember using this switch about a year earlier in training, and he used it again. All instrumentation was restored, and the mission continued. If not for John and Al, the mission would have been aborted and saving lives taken precedence.

It is a tribute to Mission Control and the crew that it took only one extra orbit to be ready to go to the moon.

Earth A question remained: was the landing system or its parachute deployment mechanism damaged by lightning strikes? No one would know before they tried to use them on their return to Earth.

Pete designated the exact location where they were to land as "Pete's Parking Lot." A precise landing was required to demonstrate this capability since it would be necessary on subsequent missions. In fact, Pete landed 580 feet short because his parking lot looked rougher than desired during their final approach.

Credit: Alan L. Bean—First Artist on Another World
Apollo 12 Crew: Commander Pete Conrad, CM Pilot Dick Gordon & LM Pilot Al Bean

Among their many activities on the Moon during their two lunar walks, Pete and Al deployed the Apollo Lunar Surface Experiments Package (ALSEP), a set of challenging and time-consuming experiments to set up properly. After the crew left, these experiments continued to measure the local seismic activity, magnet field, ion flow, heat flux, and solar wind.

Fortunately for me, the prime crew selected me to be a Capsule Communicator (Capcom) on their mission and be on their support crew. This crew did not fly or train to fly, but we did help assure that the prime crew's interests were incorporated in the design and mission planning. We also communicated directly with them in flight. I specialized in their walks on the moon and participated in other phases of the mission. This assignment was enjoyable, professionally rewarding, and helped prepare me for my own prime crew assignment on Skylab.

Once Pete and Al returned to Dick in their orbiting CSM, their Ascent Stage was detached, its engine thrusted for a short time, then it fell back down to the lunar surface. The seismometers that they had left behind registered vibrations for over an hour after it hit.

Luckily, the parachute deployment mechanism was not damaged by the lightning strike and worked as required before landing. On November 24th, they returned to Earth in their CM.

Space, Ever Farther, Ever Faster – Now!

After his career, Al focused his abundant drive on becoming an exceptional artist, a goal he certainly achieved. Because Pete and Al always wished they could have had Dick down on the lunar surface with them. In his artwork, Al achieved the next best thing as seen below.

Credit: Alan L. Bean—First Artist on Another World
The Fantasy: Pete, Dick, and Al together on the Moon.

(Note Al's two fingers up behind Dick's helmet

Apollo 12 Preflight Briefing – Ed Gibson, Pete Conrad, Al Bean, and Jack Schmitt

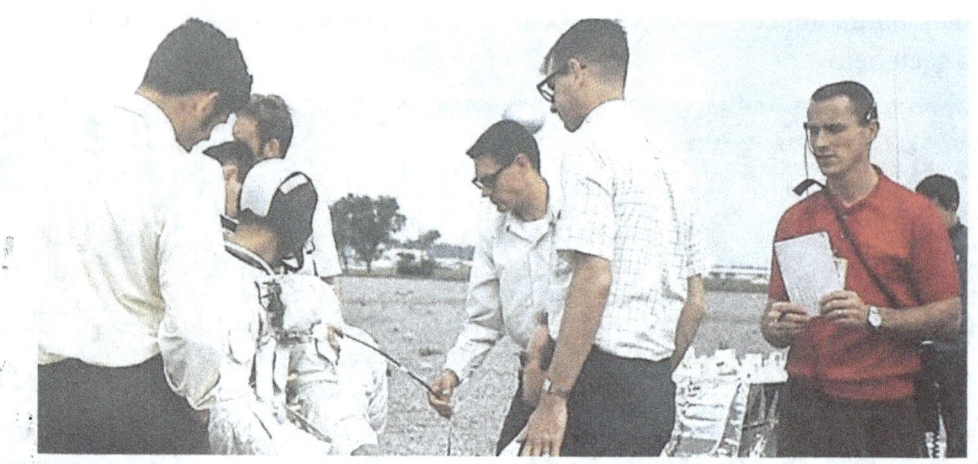

Apollo 12 Lunar Walk training at a simulated lunar site. Lunar Capcom Ed Gibson with checklist standing to the right.

Pete, Dick, and Al next to their coordinated Corvettes after training

Apollo 13 – NASA's Finest Hour

Three days before the Apollo 13 launch on April 11, 1970, Ken Mattingly, the Command Module Pilot, was replaced by Jack Swigert, his backup, because of Ken's exposure to the German measles four days previous. Since Ken was the only member of the prime crew who had not had the virus, he was the only one not immune to it. Ken was a meticulous, excellent engineer and his loss was very unfortunate for him and the mission.

However, Jack was more than ready to fill in because of his several years of technical focus on the CM development. After launch, Ken never developed measles.

Apollo 13 Crew: Cdr. Jim Lovell, CM Pilot Jack, and LM Pilot Fred Haise

The mission proceeded smoothly as the world largely ignored it—for the first two days and 7 hours. Then, after stirring fans were turned on in a Service Module (SM) Oxygen tank, an explosion occurred in the tank that abruptly changed the mission and severely threatened the lives of the crew.

Jack called down to the ground, "Houston, we've had a problem here."

They sure did have a problem, a severe one! The second oxygen tank in their Service Module did just *EXPLODE!* The source of this explosion was not fully understood until after they landed back on Earth and analysis could be performed.

It turned out that the second oxygen tank in Apollo 13's Service Module had been dropped during maintenance activities before flight. It experienced minor damage that was not evident in later inspections.

However, once filled with oxygen during ground testing, it would not empty completely. The ground personnel decided to heat the tank overnight to drive out the residual oxygen. Another failure then occurred when a surge of power in a ground high-voltage DC system caused the heater's automatic cutoff switches to fail. The tank temperature then spiked to more than 1,000 degrees Fahrenheit. Although not evident outside the tank, the insulation on wires inside the tank were damaged. Once Apollo 13 was in flight, the tank was filled with oxygen, and electricity passed through these wires, they shorted, and the tank exploded.

This complex string of failures kept the crew from exploring the Moon; that is, it caused mission failure.

From the long line of bumper-to-bumper automobile lights streaming into the Johnson Space Center in the dark morning hours, it was clear that NASA was responding with all the capabilities it could muster. Many problems had to be solved under intense time pressure and limited onboard resources. Under the exceptional leadership and inspiration of Flight Director Gene Kranz in Mission Control, every member of the Mission Control and its engineering teams clearly understood Gene's directive:

"Failure is not an option."

Gene Kranz at his Mission Control Console

In the cabin, the CO_2 level increased to dangerous levels, and the number of canisters in the LM that removed CO_2 was far from adequate. There were enough of such canisters in the CM, but, as luck would have it, they were square and not round as needed. As shown below, the crew devised a low-tech solution on the ground and utilized it.

Clearly, as every competent home handyman knows, "When the going gets tough, the tough use duct tape."

  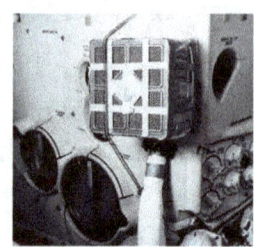

Deke Slayton holding ground solution *Jack Swigert holding the in-flight reproduction* *Solution installed and working*

The Moon's gravity was used to slingshot the crippled spacecraft around the Moon and back toward Earth, while the LM's descent engine was used to make the required orbital adjustments. When the spacecraft and its crew neared Earth and re-entry, their situation was not good. The cabin temperature had dropped to near freezing levels, Fred Haise had contracted an aggressive infection, and water had condensed on most surfaces, which created the potential for electrical shorts when the CM was again powered up.

In addition, the complex CM systems had to be brought up to be fully operational from a zero state, a procedure that was never intended to be done in flight and in a time-critical situation.

The ground team, led by John Aaron in Mission Control and Ken Mattingly in the simulator, had doggedly worked and adequately solved every associated problem. Also, partly because of the extra insulation added to the wiring after the Apollo 1 fire, no electrical shorting was observed when the CM was again powered up.

Like many others, I watched the proceedings from the Mission Control Viewing Room. And, like others, I was on the edge of my seat.

After undocking, the crew got a look at the destruction caused by the Oxygen tank explosion, which was the source of all their problems. The magnitude of the damage surprised them and most of the engineers on the ground

Damage far beyond expectations

During re-entry, one last bit of drama arose. There was concern that the explosion had also damaged the heat shield or other components required for successful and safe re-entry. The communications blackout in previous re-entries from lunar returns lasted approximately four minutes; this time it lasted for nearly six.

When the crew finally called Houston, cheers, and relief flooded through Mission Control!

A welcome sight

The splashdown in the South Pacific Ocean was an extremely welcome sight; the crew's appearance seen by people worldwide was even better.

An even more welcome sight

Appropriately, Gene Kranz called their mission "NASA's finest hour," while Jim Lovell called it a "successful failure."

Following the mission, Grumman Aerospace, which built the LM, presented a tongue-in-cheek invoice to North American Rockwell, and Whitney, and Beech Aircraft for $405,540.05 for towing the CSM most of the way to the Moon and back. However, they did provide a commercial discount. North Americans refused to pay because they had not received payment for carrying the LM Pratt to the Moon on the three previous missions.

Although the achievement was far short of what they had desired, the Apollo 13 crew holds the world record for the longest distance a manned spacecraft has reached from Earth of 248,655 miles, the distance they acquired during their swing by the Moon and headed back home.

The Apollo 13 team and NASA, in general, had been faced with extreme, time-critical adversity and came through successfully. Despite this setback, the future remained bright with NASA's broad and in-depth capability.

With leaders like Gene Kranz in Mission Control and Deke Slayton in Flight Crew Operations, NASA was ready to take on the unknown operational problems that waited in the future.

Chapter 9
Each Mission Improves Over the One Before

The next four lunar missions employed expanded hardware capabilities, longer stay times, more sophisticated experiments, and greater knowledge on the part of the planners and crew of the work to be performed. Consequently, Apollo's achievements continuously accelerated.

Apollo 14

Al Shepard Finally Flies Again

Apollo 14 Crew: CM Pilot Stu Roosa, Commander Al Shepard, and LM Pilot Ed Mitchell

Al Shepard, Stu Roosa, and Ed Mitchell launched on January 31, 1971, for a nine-day mission, the last of the "H missions" that each had two-day stays and two moonwalks. After they landed in the Fra Mauro formation, which was initially the target of Apollo 13, Al and Ed collected 94 pounds of Moon rocks during their 9.5 hours of moonwalking, Al was the only one of the Original Seven to make it to the Moon.

It was also a personal triumph for him after he had battled back from Ménière's disease that had grounded him for five years up to 1969.

Lunar rickshaw tracks leading from the Apollo 14 Descent and Ascent modules

On the way back to Earth, the crew conducted the first U.S. materials processing experiments in space.

Stu, who worked in forestry in his youth, took several hundred tree seeds on the flight. These were germinated after the return to Earth and widely distributed worldwide as commemorative "Moon Trees."

Apollo 14 CSM Landing

Apollo 15

The First of the New Apollo Capabilities

Apollo 15 Crew: Commander Dave Scott, CM Pilot Al Worden, and LM Pilot Jim Irwin

Because the J missions provided an upgraded LM capable of three-day stays on the surface and a CSM brimming with SIM bay instrumentation to survey the Moon comprehensively, Apollo 15 focused on the mission's science. Dave and Jim received exceptional training from Leon Silver, a Caltech geologist and Scientist Astronaut Jack Schmitt.

Ed Gibson

Galileo's Theory Tested on the Moon

On the final moonwalk of their mission, they explored various sites along the edge of Hadley Rille. Then Dave performed a simple yet elegant TV experiment, demonstrating Galileo's theory. He dropped a hammer and a feather together, and they landed simultaneously, which proved that all objects fall at the same rate, regardless of mass in the absence of air.

Dave proves Galileo right

At mission conclusion on the Moon, Dave parked their Rover away from the LM where he left a figurine called "Fallen Astronaut" and a plaque bearing the names of 14 American astronauts and Soviet cosmonauts who lost their lives in the pursuit of space exploration to that time.

Their landing back on Earth went well up to parachute deployment, in which only two of the three chutes deployed adequately. Fortunately, only two were required; the third one was for redundancy in situations just like they encountered. There was no doubt that this first of the new series of missions had significantly improved our performance on the Moon.

Landing with only two chutes

"After the great work of the whole Apollo Team, we did eight big walks on the Moon, so far.

Now, just six more to go.

LET'S GET WALKING!"

Ed Gibson

Apollo 16
Even Better Yet

Apollo 16 Crew: CM Pilot Kenin Mattingly, CDR John Young, and LM Pilot Charlie Duke

The Apollo 16 crew launched on April 16, 1972, for an eight-day mission, the second of the "J missions" having three-day stays on the Moon and three moonwalks.

Site Selection and Training

The Descartes area in the lunar highlands was selected as Apollo 16's landing site to allow John and Charlie to explore what was believed to be geologically older lunar material than the samples obtained in the first four landings. Between two young impact craters, it provided "natural drill holes," which penetrated through lunar regolith and left exposed older bedrock that could be easily sampled.

Like Apollo 15, John and Charlie received geological training that included several field trips. In July 1971, they went to Sudbury, Ontario, Canada, the first time Apollo astronauts had trained in Canada for a mission. Like Al Worden had for the previous mission, Ken also trained to recognize geological features from orbit by flying over earth sites in an airplane and operating the Scientific Instrument Module (SIM) like he would in space.

The Mission

They landed in an area of the lunar highlands and explored 16.6 miles on their Lunar Roving Vehicle. John and Charlie collected 211 pounds of Moon rocks during their 20.3 hours of Moonwalking spread over three EVAs during their three days on the Moon. Their exploration accomplishments exceeded those of the preceding Apollo 11 through 15 missions.

John never thought of himself as "walking on the Moon" but "working on the Moon."

Space, Ever Farther, Ever Faster – Now!

All of the time, Dave and Jim were on the lunar surface; Al studied the same surface and its environment from above with new equipment in the Service Module's SIM bay: a panoramic camera, a gamma-ray spectrometer, a mapping camera, a laser altimeter, and a mass spectrometer.

Once Dave and Jim lifted off the Moon and rejoined Al, they released a small satellite from the SIM bay to further study the plasma, particle, and magnetic field environment and the Moon's gravity. They released it an hour before their engine burn to head back to Earth. As it continued to orbit, it returned data until January 1973. The next day, on the trip home, Al performed a spacewalk in deep space, the first of its kind, to retrieve exposed film from the SM SIM Bay, which lasted 1.4 hours.

John demonstrating Rover performance.

John's jumping salute

Charlie at Plum Crater

After deployment of ALSEP and a heat flow experiment and sample collection, they drove their Rover to Plum Crater. brought back the largest rock ever retrieved on an Apollo mission, a breccia they nicknamed Big Muley after Bill Muehlberger, Principal Investigator.

Charlie in the shadow of a large boulder

The next day they drove their Rover up Stone Mountain, a 20-degree climb, and reached a cluster of five craters known as "Cinco Craters" 2.4 miles from the LM. At 500 feet above the valley floor, they were at the highest elevation above the LM of any Apollo mission and had a view that Charlie exclaimed as "spectacular." After spending almost an hour on the slope, they obtained samples from several craters and other areas. Lastly, they arrived back at a site halfway between the ALSEP and LM, where they dug a double core and conducted several penetrometer tests along a 160-foot line. This EVA lasted 7 hours and 23 minutes, the longest one of any Apollo mission so far.

Lastly, as John drove the Rover with its television camera about 300 feet from the LM so that the liftoff of their ascent stage could be televised, Charlie placed a photograph of his family and a U.S. Air Force commemorative medallion on the surface.

This EVA had lasted just 5 hours and 40 minutes.

Charlie's "Family" "left" on the Moon.

Charlie exploring far from the Rover that is visible behind him near the horizon

Damaged panels were observed from the CM on the rear of the LM after its liftoff from the Moon. Inspection revealed that the outer thermal covers were bent and torn, but the area underneath the thermal blanket was intact.

Damaged panels (Left side of the picture)

The damage occurred when gas from the ascent engine got behind these panels and lifted them off. It appeared to be a one-time occurrence, but the panels were still strengthened before the last flight.

Apollo 17
A Scientist with Operational Skills and Geological Knowledge Explores the Moon

Apollo 17 Crew: LM Pilot Jack Schmitt, Commander Gene Cernan (seated), and CM Pilot Ron Evans

Gene Cernan, Ron Evans, and Jack Schmitt, launched on December 7, 1972 for a twelve-day mission, the third of the "J" missions having three-day stays on the Moon and three moonwalks. Jack was a personable, highly competent trained geologist who joined the Astronaut Corps with me in 1965. I would have really liked to have been the scientist on this mission, but compared to Jack with his knowledge and skills, I did not know a Moon rock from a meatball.

Launch of Apollo 17 at 12:33 AM on December 7, 1972

Apollo 17 was the first night launch of an American spacecraft. It was seen directly by approximately one-half million people in the KSC vicinity. People as far south as Miami reported a red streak in the sky to the north.

The crew landed in Taurus–Littrow Valley, an area of the lunar highlands. For over 4 hours and 26 minutes, they drove their Rover 22.3 miles in their explorations and, at one point, was 4.7 miles away from their LM!

Gene and Jack collected 243 pounds of Moon rocks during their 22 hours and four minutes of explorations and stayed on the Moon for three days and three hours. Their accomplishments exceeded those of each of the preceding Apollo 11 through 16.

Site Selection, Equipment, and Training

The Taurus-Littrow site was chosen to enable the crew to obtain samples of old highland material from the remnants of a landslide event that occurred on the valley's south wall and explore relatively new, exposed material in the area.

Apollo 17's Lunar Rover carried an experiment built by Draper Laboratory at MIT, the Traverse Gravimeter Experiment (TGE). It was located on the Rover or the surface throughout each moonwalk to provide relative gravity measurements over the surface and thereby obtain information about the underlying geological structures. The 26 measurements obtained yielded productive results for the geologists back home.

Their Rover also carried an antenna that received electrical signals transmitted through the ground from a transmitter near the LM at points throughout the three EVAs. The results showed that the top mile or so of the Moon is extremely dry and most likely composed of all rock. The CSM SIM bay was packed with instruments that observed the Moon to better characterize its surface to a depth of almost a mile, its temperature, atmosphere, and spacecraft altitude to an accuracy of about 7 feet. Like on previous landings, Gene and Jack received extensive geological training, but this time Jack was not an instructor but a student and an extremely competent one. And, as Command Module Pilots had done previously, Ron trained in classrooms and aircraft.

The Mission

After the crew entered Lunar orbit, Gene and Jack landed in the Taurus-Littrow valley without significant problems and began their first EVA approximately four hours later. Early on, Gene brushed up against their Rover's right rear fender, his hammer caught under it, and the fender extension broke off. They thought it might not make a significant difference, but they found themselves covered with dust once they started driving.

Repaired" Apollo 17 Fender

Jack and Gene built a replacement fender extension by taping four photographic paper maps together, then clamping it to the remaining fender (above). It lasted through every moonwalk.

Once again, "When the going gets tough, the tough use duct tape.

On their second moonwalk, Gene and Jack collected 75 pounds of various types of rock in the valley, including the avalanche at the base of the South Massif, ejecta from Camelot crater, and, of a surprise to all, orange-colored soil at Shorty crater, which was young volcanic material (described on page below). Finally, they took seven gravimeter measurements and deployed three more explosive packages.

In the lunar desolation, Jack heads back to their Rover

To get back home, their real home, required Gene and Jack to lunar skip and bound back to their Lunar Rover, drive several miles over rocky lunar soil to their LM then liftoff in their LM upper stage to rendezvous and dock with their orbiting CSM, burn the Service Propulsion System of the CSM to head back toward Earth, re-enter the Earth's atmosphere at almost 25,000 mph, and lastly land by parachute on the ocean.

Only a minimal number of humans have ever extended themselves this far over the edge, with close to no return. Twelve, in fact, did — the American Apollo Lunar Landing Astronauts! The last two, Gene and Jack, extended themselves the farthest.

On their last moonwalk, they collected 146 pounds of lunar samples from the base of the North Massif, Sculptured Hills, and crater Van Sergand then took nine gravimeter measurements. In this last traverse, they collected a rock, a breccia, and dedicated it to several nations working in the Houston Mission Control Center.

Last two of the "Lunar Dusty Dozen"

CDR Gene Cernan

LMP Jack Schmitt

Jack next to boulder during EVA-3. Lunar Rover is visible to the right.

Lastly, they unveiled a plaque on the Lunar Descent Stage that commemorated the many achievements made during the Apollo program. Before reentering the LM, Gene said, "America's challenge of today has forged man's destiny of tomorrow. And, as we leave the Moon at Taurus-Littrow, we leave as we came, and God willing, we shall return with peace and hope for all humanity. Godspeed from the crew of Apollo 17."

After liftoff from the Moon and docking with Ron in the CSM, their ascent stage was deliberately crashed into the Moon, an impact recorded by seismometers deployed on their mission and previous Apollo missions.

As mentioned above, Jack found orange soil that consisted of volcanic glass beads, which also contained water molecules, invisible to instruments of the 1970s. It indicated the location of a volcanic vent. Recently, with the help of people familiar with extracting subtle colors from pictures, he developed a new image of the soil showing vivid hues of orange and red never seen in lunar surface pictures before. "I finally got the colors right," Jack said

Orange Lunar Soil

After leaving lunar orbit, Ron performed a successful retrieval of exposed film from the CSM's SIM bay as was done on several previous missions

SIM bay (top) on the Service Module

Ron retrieving SIM Bay film

Apollo 17's return to Earth was nominal. Also, as was the norm, the accomplishments of Apollo 17 exceeded those of each previous lunar mission. America was leading the way for all other space-faring nations. Led by America, a continuing bright future waited for all human exploration in deep space.

Space, Ever Farther, Ever Faster – Now!

Apollos 18-20
Promise of the Most Productive Missions Yet!

After the experience gained on all previous Apollo missions brought about by tremendous expenditure of funds, hard work, and personal sacrifice on the part of 400,000 American workers and their families, all elements were now in place to make the last 21% of Apollo (18, 19 and 20), the most productive missions of all.

"Yes Sir. We've got EVERYTHING FULLY UNDER CONTROL and READY!

EQUIPMENT including Saturn V boosters and spacecraft were completed or almost completed and *READY*.

CREWS of experienced astronauts, who were selected mainly for lunar missions, were set to enter final training, and were *READY*.

MISSION CONTROL was thoroughly knowledgeable and experienced in all operational procedures and was *READY*.

INDUSTRY SUPPORT was in place to deliver the few remaining hardware items, assist in their integration, and was *READY*.

INTEGRATION of all the above in knowledge and practice, the total system was efficient and *READY*.

KNOWLEDGE OF THE MOON gained on previous Apollo missions had sufficiently defined the most effective operations to answer the remaining major scientific questions and was *READY*.

DEMONSTRATION to all nations of the world that America had the resolve to finish what it started with its customary firm commitment and was *READY*.

With our near touchdown in plain sight, almost completely paid for, and all of it *READY*, how did America, our leaders, boldly move forward into the d and most productive missions

And just did, our wise, courageous national leaders decide to do?

As we in America and the people of the world watched, once America reached the 21

yard line, **our national leaders simply dropped the ball and sauntered off the field!**

WHAT?   WHY?

America failed to live up to the last three of our major mission commitments.

Without trust, how can we effectively lead others into space?

But first, before we pass too harsh a judgment, let us look at what has come out of the Apollo Program so far.

Space, Ever Farther, Ever Faster – Now!

Chapter 10
Payoffs from Apollo

Jack Schmitt before their Apollo 17 LM, Flag and Rover

What came out of it all for America? Was it just these footprints, flags, and equipment we left on the Moon?

NO!

Ed Gibson

America's immediate payoffs came right down here, at home — in AMERICA!

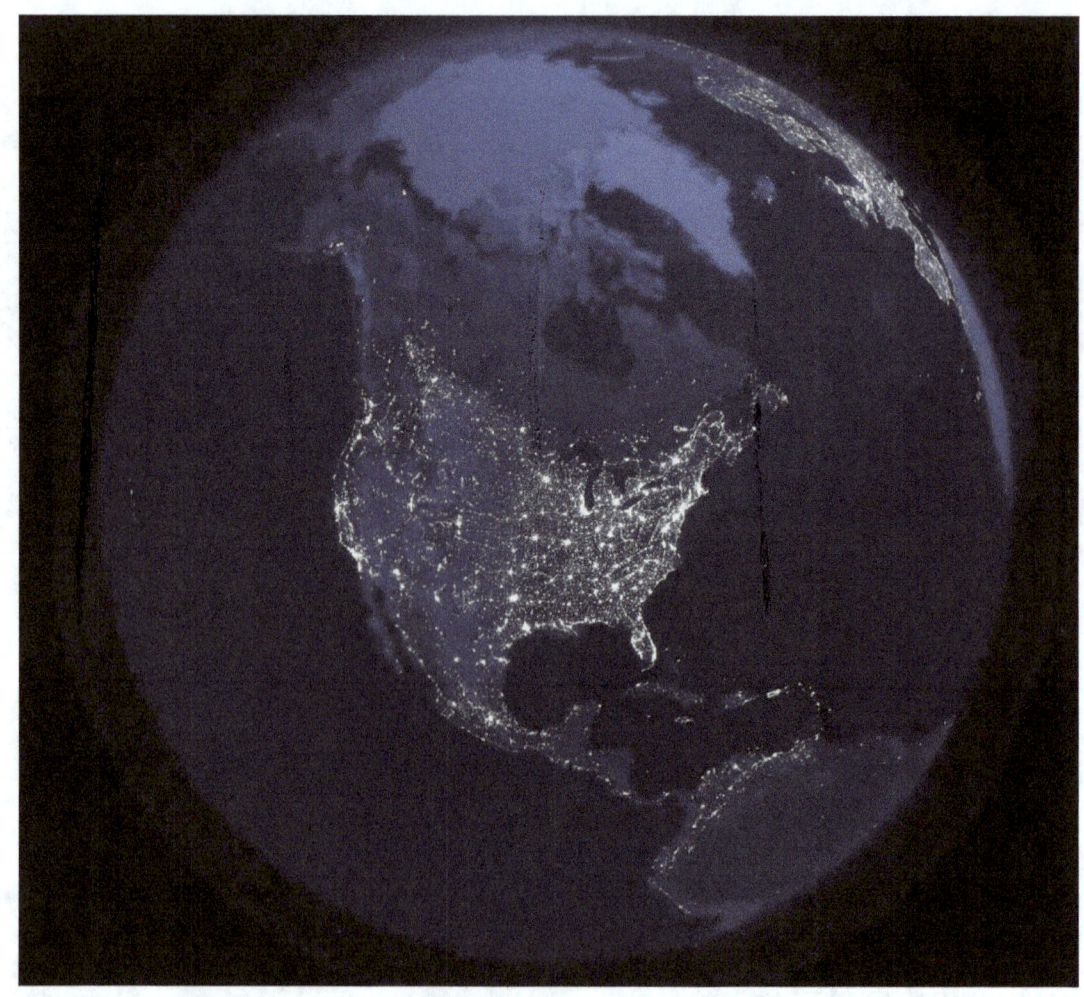

Economic and New Technology Payoffs

Along with our technology explosion came an impressive economic expansion. Results of studies by several international banks have shown that, over time, aggressive investments in new technologies yielded even far more aggressive economic payoffs.

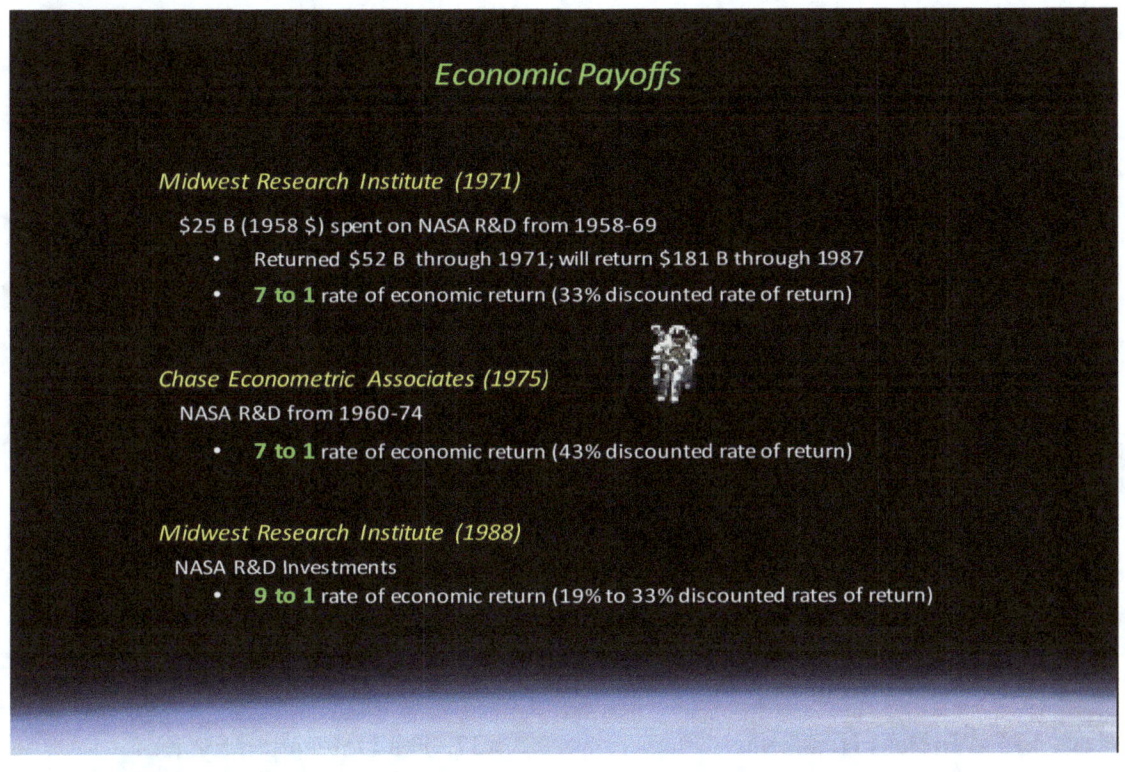

Real and undeniable economic benefits

Our first decade of space exploration and utilization, as expensive as it was, has far more than paid for itself. Why has this fact gone unnoticed, ununderstood, or unappreciated by our politicians?

On the next page, 42 inventions or new technologies coming from the Apollo Program are listed. These new technologies are truly new or significant improvements of existing technologies. It has been estimated that so far there have been over 2,000 improvements in our lives that we would not have today if they would not have come from space technology developments.

42 Inventions from the Apollo Program

1. Thermal Blankets
2. The Dustbuster
3. Advanced Cameras
4. Vacuum-sealed Food
5. Shock-absorbing sneaker Soles
6. Fireproof Firefighter Uniforms
7. Artificial Limbs
8. The Insulin Pump
9. Scratch-Resistant Lenses
10. DustBusters
11. Firefighting Equipment
12. Shock Absorbers for Buildings
13. LASIK
14. Solar Cells
15. Water Filtration
16. Wireless Headsets
17. Better Tires
18. Adjustable Smoke Detector
19. Freeze-Dried Foods
20. Invisible Braces
21. Camera Phone
22. Baby Formula
23. Lifeshears
24. CAT-Scans
25. The Grooved Pavement
26. The Memory Foam
27. Workout Machines
28. Air Purifier
29. Home Insulation
30. Infrared Ear Thermometers
31. The Portable Computer
32. LEDs
33. Ice-Resistant Airplanes
34. 3D Food Printing
35. Athletic Shoes
36. Computer Mouse
37. The Defecation System
38. The Digital Fly-by-Wire
39. NASA's Fireproof Materials
40. Cooling Spacesuits
41. Spring Tires
42. Integrated Circuits

It is clear thar each one of the above 42 items requires significant amplification to make the case for its strong positive contribution to our lives Each item is taken from and described on the website below (42 Inventions from Apollo):

https://apollo11spsace.com/42-inventions-from-apollo-program/

Education Payoffs

Lastly, underrated but most important of all over time is education itself. Space captured children's imaginations of all ages changed students' interests and career choices and led to more remarkable achievements in many technical fields.

No doubt, Apollo's children, who are now about to retire or have retired, have made our world a much better place.

Indeed, our work and accomplishments in the Apollo era provided a broad spectrum of payoffs in technology, economics, and education for America. Further aggressive reaches into space can give a host of equivalent profits for our Nation and its people. All it takes is leadership with a solid vision to see the opportunities and needs right before us, courage to vigorously pursue them, and commitment to stay with it until fruitful results are totally realized.

Can we afford not to take advantage of this technology, economic, and education stimulus for the good of the American people? Our enhanced abilities and benefits have been clearly demonstrated. We should not doubt it but recognize it and use it!

Achievements in space allow all Americans to celebrate and reflect on what we, a free people with capability, spirit, and determination, can achieve. With each such event we celebrated, like Americans did at many locations across our Nation when our soldiers paraded home at the end of World War II. Fortunately, we were able to do the same now as some space explorers returned.

America's pride on full display - Apollo 11 homecoming parade in New York City

Each generation deserves the stimulus and right to take pride in the accomplishments of our freedom-loving Nation.

It is and hard to conceive, let alone believe, that the tremendous capabilities that we developed and the benefits we received were swept away by a new set of leaders who saw but did not understand and heard but could not appreciate.

> **At this point everything is in doubt. We did it before…**
>
> *but with our degraded attitude, can we ever do it again?*

It is regrettable that in addition to the excitement and apparent accomplishments of the Mercury, Gemini, and Apollo programs, it went largely unrecognized that they were also essential factors in ending the Cold War. Meeting their enormous technological and organizational challenges resulted in significant advancements in science and engineering, enabling earlier and greater progress in a wide variety of fields of high value to humankind.

Chapter 11
Our Depressing Plunge

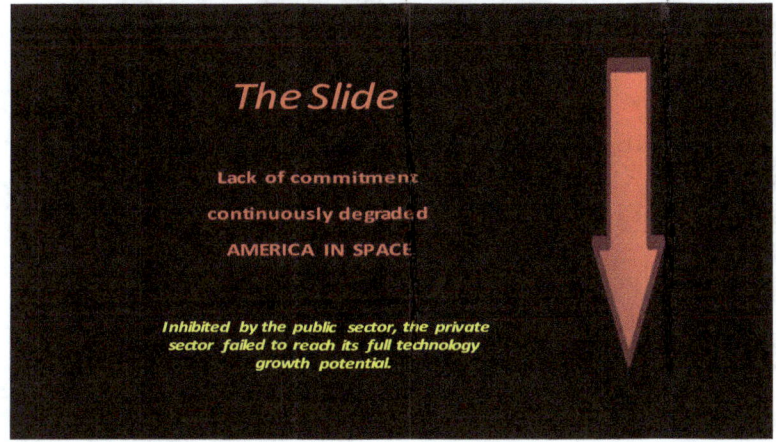

Hang on! It's free-fall right to the BOTTOM!

Our coaster of accomplishment had climbed to the top of the first hill. Now, brace yourself; here comes the plunge.

The public sector, that is, the President and Congress, no longer saw the value in an aggressive space exploration program; any exploration by human flight was again restricted to Earth orbit only, and the pace of new initiatives in space faded.

The private sector, no longer challenged as they were in Apollo and before, could no longer keep up the vigorous growth of innovative technologies.

Ed Gibson

SUMMARY OF AMERICA IN SPACE

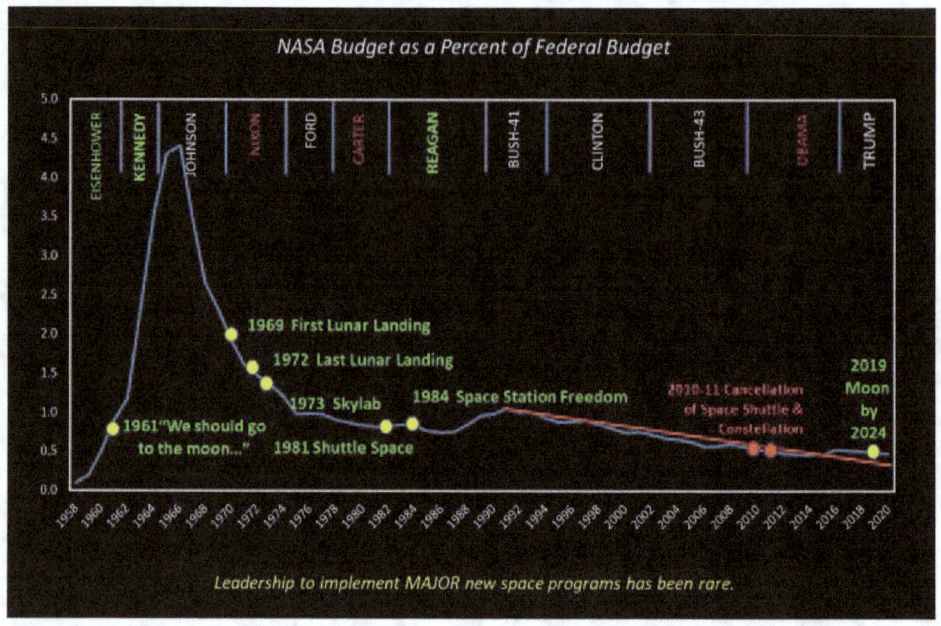

The numbers tell it all!

It's clear that after the initial impulse, the erosion has been steady except for one short period under President Reagan, who had the leadership to assure the implementation of the Space Shuttle and the Space Station. But in general, aggressive action in Vision, Courage, and Commitment at the national level of space leadership was absent.

With only a first term, President Trump was on track to ensuring our return to the Moon by 2024 and n moving on to Mars, which might have put him standing next to President Reagan. However, he did not get an immediate second term in a controversial election. Yes, there was light at the end of the tunnel. But. as we have learned, politicians can snuff it out or just sneak in to slip in more tunnel and delay progress!

So far, Human has been driven into space by national competition rather than the inherent value of being a space borne civilization. Will the American people's vision, courage, and commitment prevail, or will the political interests of our national leaders again take precedence?

Leadership

As later events have demonstrated, this radical change in direction was a precursor of things to come and an indicator of a severe flaw in our Nation's decision-making processes, which has affected how aggressively We, the Human, Enters

Space, Ever Farther, Ever Faster – Now!

Space. In fact, the seeds of a new vacillating political behavior were planted as early as our first lunar landing.

There is no doubt that our human space program was initially motivated by international competition with the Soviet Union, that is, the need to demonstrate technical superiority for our own national security. It was a race to the Moon, and we ran it like a coach once told me how to run the quarter-mile dash. "Son, you sprint the first 100 yards, then gradually increase your pace."

Obviously, once the race was over, it was time for a breather, a pause to reflect. And when we did, we realized that the race was not over but had shifted to a never-ending marathon motivated by our own drive to continually grow, refine, and utilize our understanding of the physical world in which we have found ourselves. Additionally, we are driven by our growing commercial interests, and the continuing drive for technical superiority and security. But what happened? How did it all go so wrong?

In January 1970, after Apollo 12 concluded, NASA canceled Apollo 20 so that its Saturn V would be available to launch the upcoming Skylab Space Station fully outfitted on the ground. instead of trying to construct it in flight within the used Saturn, IB spent the upper stage. A spare Saturn V booster was unavailable because of budget restrictions limited Saturn V production to 15. The funding pendulum swung from high positive to high negative without even a short pause in between.

In April 1970, after the Apollo 13 incident, NASA canceled Apollo 19, citing budget cuts.

Then in September 1970, NASA also canceled Apollo 18, citing more budget cuts. President Nixon even proposed canceling Apollo 16 and 17, but fortunately, he was talked out of it.

Budget cuts were how America truncated its planned explorations of the Moon. These decisions made in the early 1970's greatly weakened how America has entered space. But what were the specific motivations behind them? Many factors have been cited, including: (Please see the following page.)

MOTIVATIONS

Factor	Budget Cut Motivations and Excuses	Actions of a Strong America Entering Space
Future Needs	Skylab required a Saturn V booster.	Don't steal from the future to fix the shortfalls of today. Fight for what is needed today, today.
Too Dangerous	After Apollo 13, some politicians finally realized that spaceflight was dangerous, finally.	After fixing or accepting all identified safety issues, fly. If that is not acceptable, get out of the business.
Political Gain	President Nixon feared a tragedy in space would damage his chances for re-election.	Politicians are in office to do what is right and serve the American public, not the other way around.
Public Vacillation	Some thought that America's public interest in space had substantially decreased.	Only huge public opinion swings should deter a program from its goal, which was not the case from 1965 to 1975.
Building vs. Using	To build the new, commitments to utilize the old were dismissed entirely.	Resist the temptation to jump to new ideas. Demand discipline to fully utilize past investments.

This failure to complete what was already planned, in place, almost all paid for, and taken as a commitment by those invested in our space effort at home and abroad, has been one of our most discouraging losses of nerve and commitment in American history!

In this wasteful cancellation of the last three Apollo missions, much hardware was sent off to museums, including:

1. Two Saturn V boosters are on display at the Johnson Space Center (JSC) and the Kennedy Space Center (KSC) Apollo-Saturn V Center, except that the first stage of the Apollo 19 Saturn V is on display in Pearlington, MS.
2. Two unused Lunar Modules are on display at the National Air and Space Museum in Washington, DC, and at the KSC Apollo-Saturn V Center in Florida.
3. One unused CSM is on top of the Saturn V display at JSC.
4. One Saturn IB, the Skylab Rescue vehicle, which later became the Apollo Soyuz Test Project (ASTP) unused rescue vehicle, is displayed at the KSC Visitor Center.
5. One of the Saturn V third stages became part of Skylab as a backup Orbital Workshop and is now on display at the National Air and Space Museum.
6. Other hardware was used elsewhere or scrapped.

Spacecraft were NOT made to sit in museums — They were MADE TO FLY!

President Kennedy exhibited outstanding leadership in initiating America's vigorous space program despite the ongoing cold war, increasingly hot Vietnam war, and the fight for racial equality at home. President Johnson continued it while also initiating his War on Poverty. But then, President Nixon put on the brakes when he canceled the planned and already-paid-for remaining missions that confined Americans to low Earth orbit for over the next half-century. Simply a wasteful retreat! As an infant toddler, America finally stepped into Space, then had to turn around and step right back!

Human wonders...

"HEY! I Just started!

But now I'm grounded and back where I began. Just 3 steps forward, now 3 steps back!

WHERE to now?!"

Chapter 12
Skylab

Further exploration of the Moon is now out; however, we can still utilize the earth's orbit. But how can we ever follow the excitement and fulfillment of lunar exploration?

The Long Tedious Road to a Real Space Station

The seeds of Skylab were planted about 2.4 centuries ago when Edward Hale, in 1869, wrote "The Brick Moon" mentioning a space station. Over many decades in the 20th century, many science fiction writers imagined laboratories in Earth orbit with their inhabitants floating inside without gravity. More practical minds took over from the romantics once they realized that the ability to fly in space and its many real benefits were within reach.

Although greatly discouraged by America's withdrawal from vigorous outward space exploration, a good segment of the aerospace community understood that there was still much to achieve within the near-Earth space already secured. Their sentiments were: "We should explore and utilize this new environment to advance our science and technology."

The opportunities for further exploration and its utilization were clearly described by Wernher von Braun in several articles for the public in Colliers Magazine even before the Apollo Program was conceived.

Von Braun was a visionary, engineer, and an inspirational leader who brought clarity and specifics to the discussion.

Of course, we should not forget his Nitze past. But we also should not ignore his strong, humane leadership or his substantial positive contributions to our American space program.

Wernher von Braun

It was first thought that the on-orbit construction of the space station inside the Hydrogen tank of a spent Apollo upper stage was cheap and would suffice. This "wet workshop" received considerable study. Scientific experiments and instruments were added to the study, including seven high-powered solar telescopes already in development for Apollo, integrated and called the Apollo Telescope Mount (ATM).

In 1959 von Braun proposed using the third and final stage of a Saturn launch vehicle that could be turned into a space station on orbit. The idea did not immediately catch hold, but four years later, NASA testified before congress that a manned Earth-orbital station should be considered in NASA's future.

Wernher von Braun and his team again proposed their idea in an advanced study. In time, it was realized that launching Skylab fully outfitted on a Saturn V would create a much better facility, save crew time and, in the long term, be much more cost-effective. Finally, by 1973 the total Skylab system was ready for launch.

The launch went well for the first 63 seconds. Then problems started when the micrometeoroid shield around was Skylab ripped off. The shield shown was intended to be above (the label is on the extreme right halfway down). Once deployed in orbit, it would form a cylindrical sheet of metal wrapped around the large cylindrical body of the Saturn Workshop, which would accommodate the crew. They would live, work, and stand a few inches away from it to stop or break up any micrometeoroids that might otherwise penetrate the pressurized living quarters.

Unfortunately, when going through approximately Mach 1 in the launch, some high-pressure gas leaked under this shield, lifting it into the high-speed airflow, ripping it off, and leaving it to fall into the Atlantic Ocean and reside at its bottom for the coming centuries.

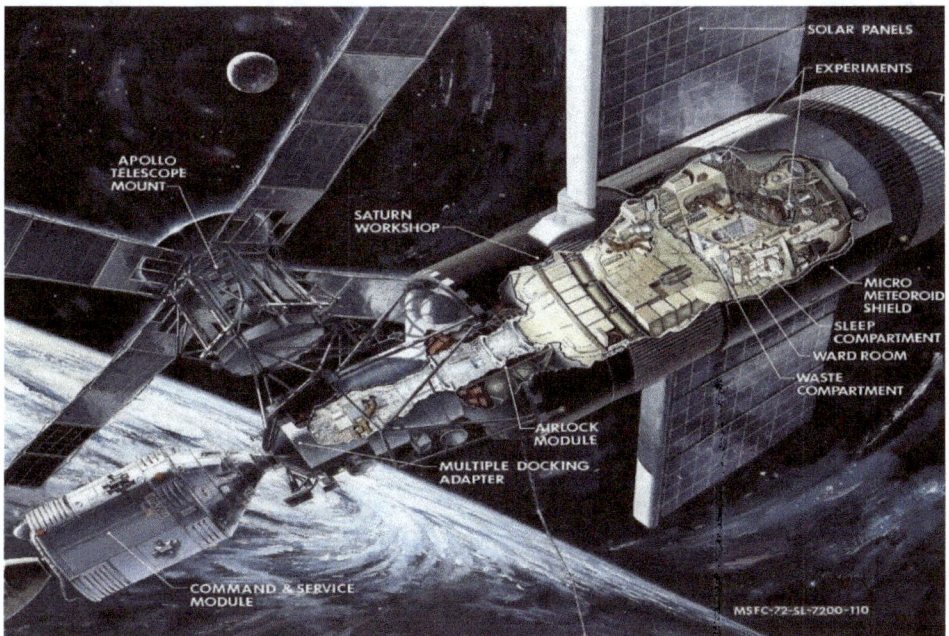

This loss of the station's micrometeoroid and solar heating protection were just one part the station's growing set of problems.

When a strap from the shield wrapped around one of the two undeployed solar panels, it prevented the panel from opening once in orbit. The other solar panel completely ripped off later in the launch and headed for the ocean. Immediately there was a shortage of power. Second, the shield was designed to reflect sunlight and provide thermal protection. Thus, the station was severely underpowered and overheated once in orbit.

The launch of the first crew, which should have launched the next day, was delayed ten days to come up with a plan and equipment to address the high-temperature problem first and the stuck solar panel second.

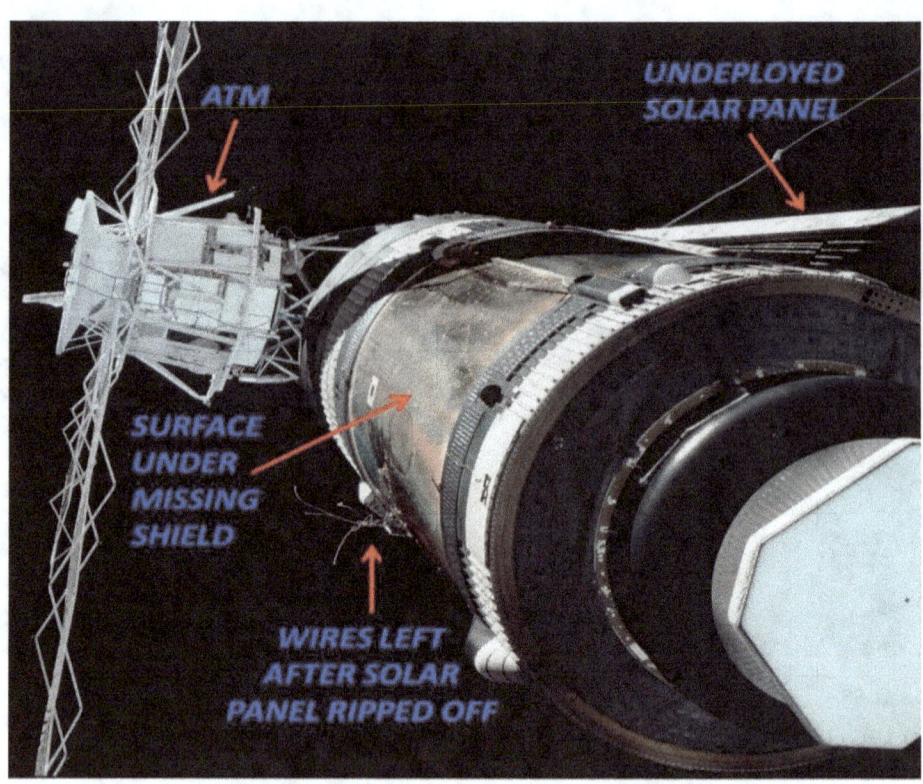

Skylab, as viewed by the first crew at mission start.

The numbering system for the Skylab flights was confused well before liftoff. Initially, the launch of the space station was Skylab 1, followed by the three manned missions of Skylab 2, 3 and 4. It was decided late in the game to only number the manned missions, Skylab I, II, and III. This latter system is employed in this book, although both are still used in the general literature. On our patch, "Skylab 3" is used, which is not valid in any system but could not be changed because our clothing was already launched on the station. Unfortunately, all of this has continued to cause much confusion.

The first, second, and third Skylab crews

Skylab I:
CDR Pete Conrad,
Science Pilot Joe Kerwin,
and Pilot Paul Weitz

Skylab II:
Science Pilot Owen Garriott,
Pilot Jack Lousma,
and CDR Al Bean

Skylab III:
CDR Jerry Carr,
Science Pilot Ed Gibson
and Pilot Bill Pogue

Skylab I

On May 25, 1973, the first crew launched and rendezvoused with Skylab. The picture two pages back shows what they immediately encountered: the remaining solar panel was only slightly opened, and a strap from the departed micrometeoroid shield prevented it from opening farther.

Before docking, Paul, in a stand-up EVA, tried to free the arm by pulling on it from the open hatch of the CM, and the strap and arm remained fixed. They docked with Skylab after solving a complex problem with the docking latches that required them to depressurize the CM again.

The following day when they opened the hatch into Skylab, they were greeted with a temperature of about 130°F. Before launch, we reassured them, "No worry, guys, just like Arizona, it's a dry heat." They were not reassured. Immediately, they erected a solar shade through a scientific airlock on the sun side of the station that, once outside, opened like a huge parasol. After a few adjustments, it mostly worked. The temperature inside dropped by about 40°F to 90°F.

Credit: NASA/MSFC-73-7200-2
The undeployed solar panel, as viewed by Pete, Joe, and Paul

When they began activating Skylab systems and experiments, none of the crew experienced debilitating space sickness, but they found the pace of the work very demanding.

Freeing the solar panel, which was nearly closed and held down by a strap from the departed micrometeoroid shield, would be far more complicated than bringing down the temperature. The strap had to be cut, and the solar panel arm rotated around the hinge to a 90-degree angle to the side of the Skylab Workshop so its solar panels could be deployed.

On the ground, a team headed by Rusty Schweickart worked on EVA procedures to erect the solar panel. Rusty, Story Musgrave, I, and others worked in the MSFC water tank for many days, developing strategies that we thought would work. They involved using a 25-foot-long cable cutter to cut the strap and a long tether to pull the arm up once it was free of the strap.

Rusty Schweickart and Ed Gibson in the MSFC water tank developing procedures to help save Skylab on the first mission.

Finally, on day 14 of the mission, the EVA to repair Skylab began. The procedures developed in the water tank did not work as well in space partially because the crew could not stabilize themselves sufficiently without handholds or footholds. They struggled, but the cable resisted. Pete came back along the arm to examine the position of the cutter when it finally worked. The arm jerked up to 20° open, and Pete flew out to the end of his umbilical. Unfortunately, the fluid in the damper designed to slow the erection under normal conditions had frozen solid and more work would be required.

Hence, more struggling.

Finally, Pete hooked one end of their rope to the free end of the arm, hooked the other end next to the hinge, put the middle of the tether over his shoulder, and pushed up. Nothing. Then, when both crewmen pushed up in unison, they felt a hard snap and went flying out to the end of their umbilicals, Pete, for the second time. When they turned around, they saw the arm becoming fully erect.

Fortunately, what had been feared by Rocco Petrone and others on the ground did not happen; the umbilical, which carried the crewman's oxygen, electrical power, cooling water, and communications, did not get crushed in the hinge at the base of the panel as it rotated open.

Immediately the power flow began a quick rise. Skylab I now had enough power to complete its mission and, as time would show, the following two missions as well. Everyone in Skylab, especially the last two crews felt tremendous gratitude for Pete and his crew's work to save the Skylab space station.

Pete's crew completed observational goals and a high percentage of their preplanned earth and sun observations. In fact, Paul obtained good data on a high magnitude solar flare that lasted for many minutes. In the development and repair of Skylab, the qualities that America employed to make our lunar landings so successful were again amply demonstrated: Vision, Courage, and Commitment.

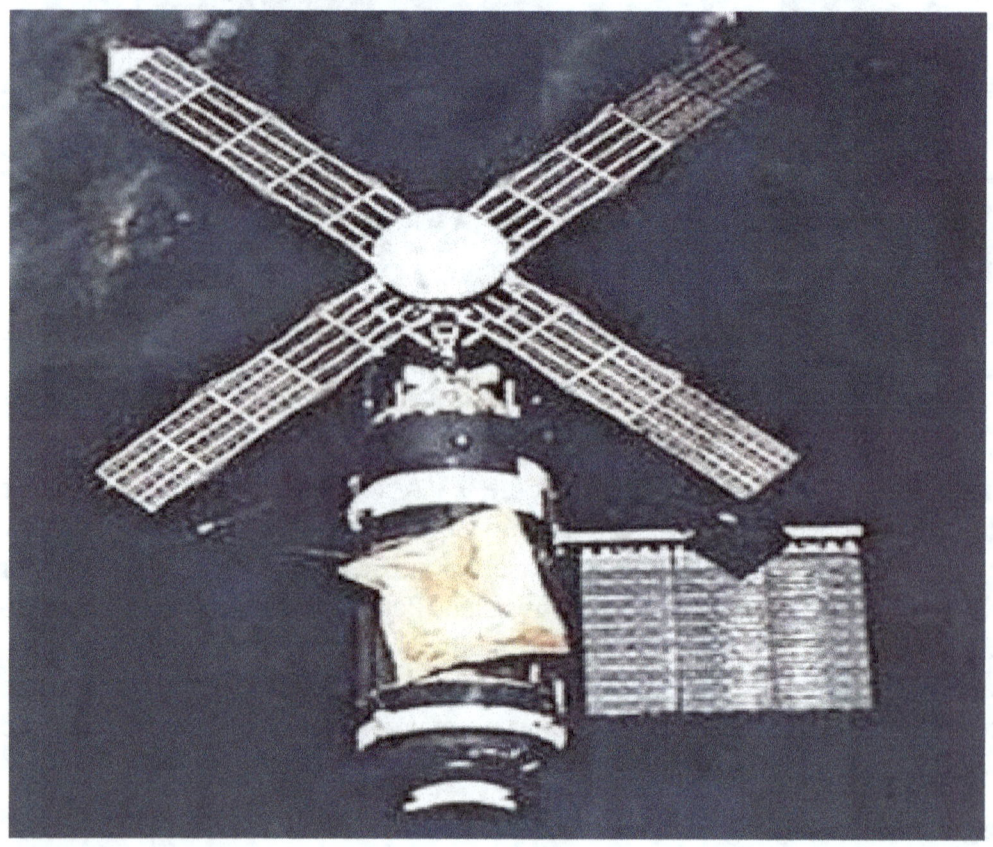

Skylab as viewed by the first crew at mission end.

The Crew's excellent work to erect the solar panel (4 pages previous) and then the parasol sunshade (center) saved their mission and the Skylab Program.

Skylab II

Science Pilot Owen Garriott, Pilot Jack Lousma and CDR Alan Bean

Because of worries over the ultraviolet deterioration of the deployed makeshift parasol shield and degradation of its rate gyros, the Skylab II crew launched three weeks early on July 28, 1973, at 7:10 AM. Before launch, the mission's duration was changed from 56 to 59 days to enable a more routine recovery operation at landing.

The crew on the pad before launch was so relaxed, in fact, that Jack dozed off a bit. However, only one hour after liftoff, Jack experienced nausea. A bit later, the other two crewmen experienced the same problem but to a less degree.

The equipment transfers power-up and reactivation of the space station took longer than planned because of the added systems and hardware troubleshooting, CSM reliability concerns, and, unfortunately, increasing crew motion sickness, especially Jack. He made it through the rendezvous and docking using scopedex, the anti-motion sickness drug that works well on the ground but not too well in space. The doctors and we, the nine Skylab crewman, chose scopolamine-dextroamphetamine (Scopedex), an upper and downer combination.

The fact that Jack experienced the worst symptoms of all came as a surprise. Jack was a Marine, capable and determined individual who had a very high resistance to motion sickness (on the ground!)

But space sickness was proving to be a different type of malady. Jack got very ill after his evening meal while Al and Owen experienced some stomach awareness for a half-day but could keep it under control by moving slowly. This sickness was unexpected because the Skylab I crew experienced only minimal nausea, and Al did not experience any symptoms when he flew on Apollo 12. Fortunately, their resulting in-flight adjustments to each activity rate worked out reasonably.

However, Al realized that they could not work at the pace that mission control had laid out for them. Instead, he suggested that they take some time off to rest in their seep compartments, slow down the projected pace of work, postpone their EVA planned for their fourth day, and shift their day off planned for a week into the mission. The doctors suggested that the crew might feel better if they took more Scopedex and performed several series of tipping their heads side-to-side forty times for 10 minutes each, which was a bad idea. Rather than helping, the head motions aggravated their nausea.

After sunshade erection, the film was installed in the solar telescopes, experiment samples were deployed, and a search was performed for leakage from a Reaction Control System (RCS) jet quad. Although planned for 3.5 hours, the EVA had lasted three hours longer. Fellow passengers on Skylab II also had to adjust. Although it took a few weeks of thrashing legs and spinning in circles, a standard run-of-the-` spider got to work and built a very credible web.

Unfortunately, the NASA space station environment did not include insects, so this spider could not take comfort for long in knowing that it was one of the first spiders ever to become a space traveler.

Spider in its home on Skylab II

Bruce McCandless had done a great job helping to develop the manned maneuvering unit (MMU) he and was the first crewman to fly it outside during a Space Shuttle flight. This testing was essential to assuring future crew safety.

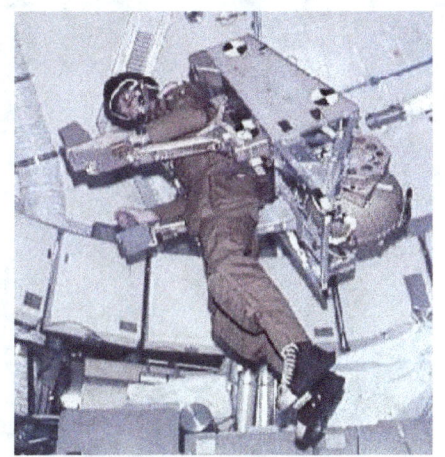

Al testing Manned Maneuvering Unit

Jack cleaning air circulation system intake

If an object were loose, it would go with the airflow and end up on this air intake screen as I did when I tried floating free during sleep. This screen soon became our "Skylab Lost and Found Department."

Skylab III

CDR Jerry Carr, Science Pilot Ed Gibson, and Pilot Bill Pogue

No doubt I was in fast company. Commander Jerry Carr was a *"United States Marine Corp Aviator"*, which says it all. Pilot Bill Pogue was an Air Force Fighter Pilot, who flew combat missions in Korea, was a member of the Thunderbirds Flight Demonstration Team, and a Test Pilot. I flunked first and fourth grades but played high school and college football and ran track. And eventually, I got good at school. Maybe if I worked especially hard, I could also more quickly get good at being an astronaut.

We had little access to the high-fidelity training facilities and simulators until Skylab II launched. The first and second crews naturally had rights to all these premier facilities until they flew. Once they did, training became an all-out effort. In particular, the Commander and Pilot had to be at their best to safely land the Shuttle after an abort, if required.

Lastly, many people realized that we would be their last opportunity to get additional equipment and new experiments up to and used on Skylab. The train would soon be pulling out, and no one with new and additional equipment wanted to miss it.

Many new experiments and some new equipment were put on board at the last minute. We had little training and experience in their operation until after launch.

Also, Dr. Lubos, in 1973 had discovered a comet while working in his Observatory in Hamburg, Germany. It promised to be very bright as it went around the Sun near the end of our mission.

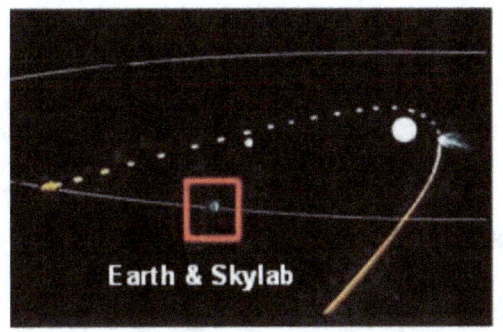

The trajectory of Comet Kohoutek in late Dec. 1973 circling the Sun and Earth orbits

This was to be a "once-in-every-2000-years" comet that some even speculated was last identified as the Christmas comet at Christ's birth. Indeed, this opportunity took high priority.

We also added to our workload when we asked Ken Kleinknecht, our very supportive Skylab Manager at JSC, to add significantly more training in Earth Observations so that we could be much more knowledgeable and competent in taking hand-held camera data as we went over targets of opportunity on Earth and added our descriptions. He obliged, and we received 40 hours of additional briefings from some very competent Earth scientists and additional film to use in flight.

For sure, great support!

Of course, all this new hardware and tasks required sufficiently more training and installation time and in-flight operations time. Thus, many new experiments and new equipment were put on board at the last minute. We had little training and experience in their operation until after launch. But, if the guys before us could do it, so could we.

Of course!

All this new hardware and tasks required sufficiently more training and installation time and in-flight operations time for us to perform efficiently, accurately, and comprehensively. But no problem.

Of course!

The Skylab II crew had ended up (after a slow start) at a very high-performance rate and achieved 150% of what they were scheduled to do. And, of course, we could start at close to Skylab II's ending pace (28 man-hours of work on equipment per day), and everybody would be happy.

Of course!

Lastly, a few weeks before launch, many in NASA, including the medical community, joined the heightened concern over space sickness.

Our CM was loaded to the hilt with the added items needed to repair, replace, or upgrade equipment as we approached launch. (Please see the sketch of our CM below.) Additional cameras and film, the replenishment of supplies because of added planned usage, 160 pounds of additional food for the lengthened stay, and last but not least, 500 gypsy moth eggs.

Padding typically used to cushion the stowed items was replaced with some of our additional clothing. All these new items would have to be unstowed and restored properly once we entered Skylab and began working.

We also wished to come back after 12 weeks in space in excellent condition in appearance and in fact. The second crew bumped it up to one hour a day and felt that was still not enough. Therefore, we elected to exercise for one and a half hours a day. It would crowd our schedule even more, but we thought, along with our medics, the results would be worth it and we could do it.

Of course!

In Congress and NASA worry over space sickness was growing. Funding of programs after Skylab were in question if astronauts would be sick for most of a mission. Therefore, space sickness was a growing and major concern for America in Space!

The message delivered to us was clear. "You and we will do whatever is required so that you **_DO NOT GET SICK!_**" We understood the need and were determined to do our best not to get sick but also believed that no one understood the issue well enough to know just what basic precautions we should take. But we will do it, whatever it is.

Of course!

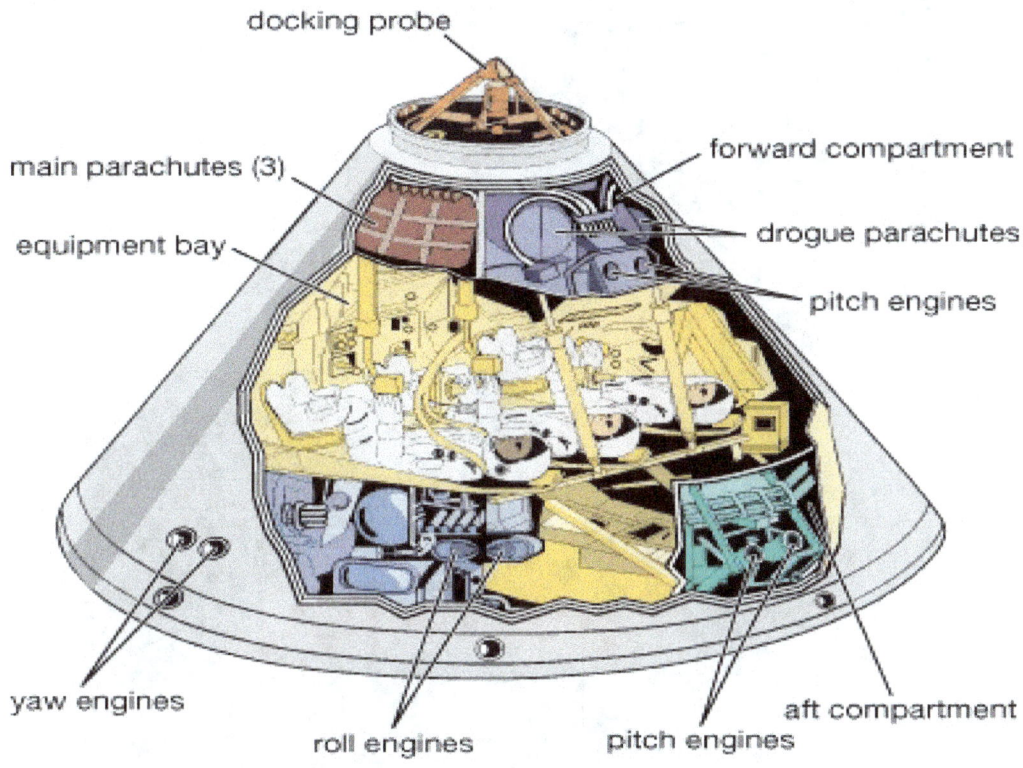

Sketch of our Command Module

An experiment that each of us was to conduct in flight and on the ground before and after the flight was the Rotating Chair (top of next page). Sitting in the chair with our eyes closed and the chair rotating at many revolutions per minute (up to about 25 rpm), we made head motions (up/down, left/right, and rocking shoulder/to shoulder) to an audible cadence.

This vestibular experiment was a good predictor of our resistance to motion sickness on the ground. Eventually, we would start to experience very identifiable symptoms of illness such as sweating palms, the appearance of skin pallor, stomach awareness, dizziness, and other sets of symptoms unique to each of us.

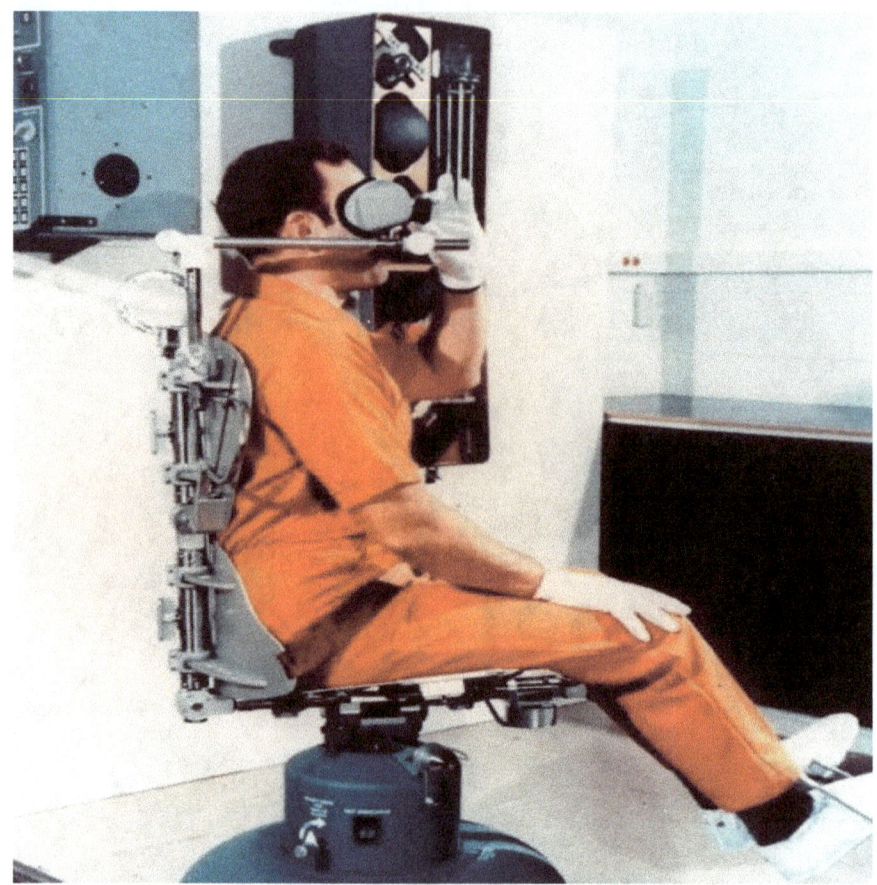

Rotating Chair

The maximum rotation rate that Jerry or I could tolerate was about the same as most other Skylab crewmen. However, we joked that we could put Bill in the chair spinning at maxrate late in the day, and when we came back in the morning, he would still be spinning around and going strong. We could not make Bill, the Thunderbird pilot, old lead ear, sick!

Also, in the weeks before launch, we each took the added precaution of flying several acrobatic missions in the T-38 aircraft. It had a max roll rate of two revolutions per second and, in a parabolic maneuver, over 18 revolutions could be achieved before a pull-up was required. It sure tumbled our internal gyros but made us even more resistant to motion sickness (in an aircraft).

Ed, Jerry, and Bill next to a T-38 *NASA T-38 in flight*

However, the politicians and some medics were not satisfied with our well-calibrated resistance to motion sickness and like in Skylab II, insisted that we also take some medication. It gave us slightly increased resistance to motion sickness on the ground but made us feel a bit dizzy. I did not think it was wise to start our mission off in that condition, nor did I feel good after taking it. Jerry felt the same.

But Jerry and only Jerry who would fly an abort after launch, if required, got a dispensation, and did not have to take it until after we docked with Skylab. Although we each had little say in the matter, we did not believe taking Scopedex before the flight, which had only an unknown advantage, if any, was the right thing to do. But we did as we were "*strongly*" advised to do and took our medicine.

Clearly, we did not like being put in this situation and had lost confidence in the fairness and rationality of the decision making behind it.

Our booster began the first, slowest leg of its trip, crawling out to the launch pad.

As launch day approached, each of us became more excited. It would be an excellent opportunity to add to the significant new knowledge about how humans can survive and thrive during almost three months in space, return mountains of top-quality data in Earth and solar observations and contribute in many auxiliary areas. We each had broad smiles as we discussed it all before the flight.

Two days before the countdown began, inspectors identified stress cracks in all eight fins of our Saturn-1B booster. The launch was slipped by five days as a precaution while the fins were replaced. Additional cracks were found in seven of the eight S-1B/S-IVB interstage beams, which required an extra day to fix. With no ill intent and a smile on our faces, we began calling our booster old Humpty Dumpty. Apparently, the good troops working so hard around the clock to get the booster ready in time did not really appreciate our "humor," but much to their credit, they said nothing… yet.

At last, on November 16, a crisp, cool, clear day, the whole Skylab III team was ready to go. We had breakfast with Bill Schneider, head of Skylab at NASA Headquarters; Ken Kleinknecht, head of Skylab at JSC; Deke Slayton, Head of JSC Flight Operations and Al Shepard, Head of the Astronaut Office.

Note that the lady serving us (right) was wearing a mask. That was because we had become NASA's very expensive laboratory animals. They wanted to make sure that if we got sick in space, it was not from bacteria or virus we that brought up with us from Earth.

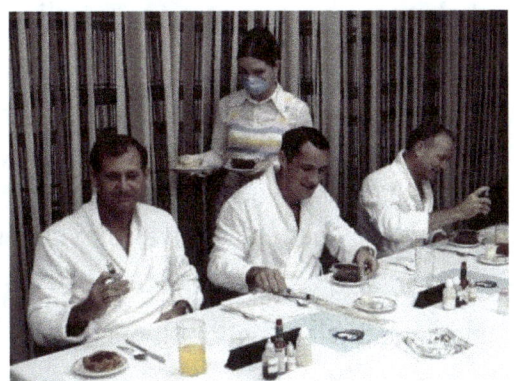

Breakfast before launch the syringes contained just` a salt-water solution.

When we arrived in the suiting facility, Joe Schmidt took control of the process just as he had for everyone to fly since Al Shepard flew in 1961. His expertise, friendliness and careful mothering of every crewman put everyone at ease.

My brief encounter with Deke Slayton after suiting up and on the way to the launch pad.

Although it might look like it, he did not say, "Bless you, my son."

Finally, the three of us and Al Shepard climbed into a van and headed out to Launch Complex 39B. As we rode, Al's big piercing blue eyes bored into each of us, looking for any sign that one or more of these rookies were not ready to go. Soundless, I stared back with a defiant smile, "Not you, Big Al, or anyone else is taking my seat!" Soon we arrived at a 37-story building. We took an elevator to the top floor and walked to the end of a long hallway where I waited outside to enter our office that now, with all the added storage, appeared to be a tiny room. Because I sat in the center seat, I was the last to enter and had about 20 minutes to just stand outside on the gantry and soak it all in.

No doubt this time was different than any of our practice runs. As I leaned against the surrounding structure, I could feel it popping and creaking and groaning under the weight and frigid temperatures of a million pounds of liquid Hydrogen and liquid

Oxygen freshly loaded into the building below. As I saw lights flashing, computers blinking and gases venting, I realized that this cold, lifeless building was finally coming alive!

Twenty minutes before launch, we got a message from those who had worked so long and hard to get the booster's cracks repaired and the booster ready to go in time. "Good luck and God Speed, from all the King's horses and all the King's men." They got us! They had the last word, which gave us a good chuckle.

Liftoff is an exciting time, and any crewperson who is not excited does not understand what is about to happen.

Finally, Bob Crippen, our steady, reliable Capcom, started counting backward from ten. We then heard a tremendous sucking sound as propellants were ripped into combustion chambers, a noise that Bill said sounded like they had just simultaneously flushed every toilet in the Houston Astrodome. Then, far below and in less than a second, we felt eight rocket engines ignite in a ripple fire, and we crept off the pad at one minute after 10 AM.

The front of my mind was intently focused on gauges, computers, and potential abort procedures, even as a bit of whisper escaped from the back of my mind, "Hey you, the basement of your building just exploded!"

At around one minute into the flight, we went through the speed of sound and hit the maximum of the aerodynamic forces and turbulence that built as we rammed through the mounting wall of air resistance ahead of us. The vibration became severe. I felt like a mouse strapped to a jackhammer.

Liftoff!

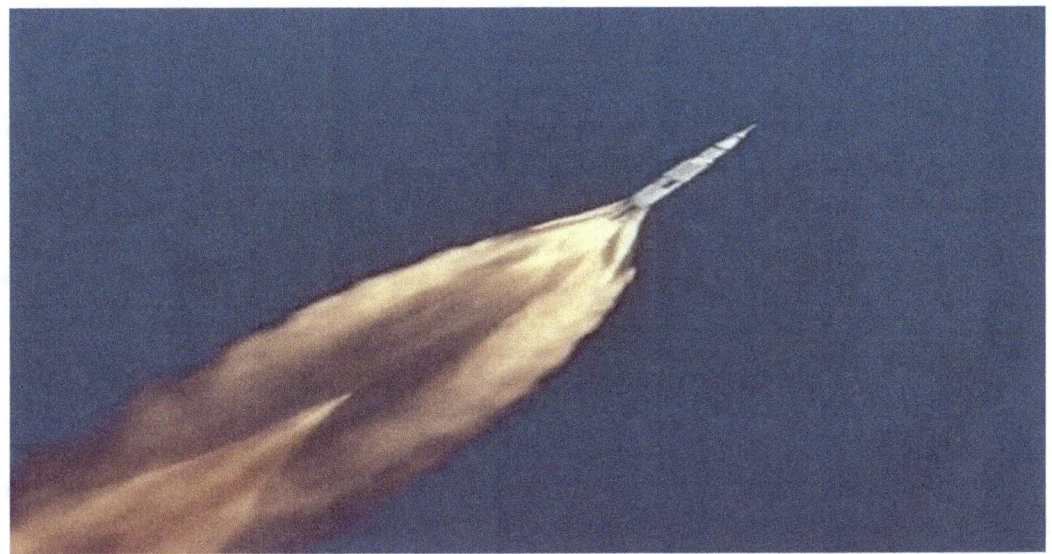

After staging, which felt like a head-on crash quickly followed by a second impact from the rear, I glanced out the window and saw the surface of the Atlantic Ocean and the curved horizon. After what seemed to be only a few minutes and a few switch-throws later, Italy was sliding underneath at smooth five miles a second.

Italy seen from Skylab III

In an unexpectedly short time, we rendezvoused and docked with Skylab. Eventually, we entered the station and immediately started to relocate some stowage.

Soon Bill said, "I'm not really feeling too terribly well."

Jerry replied, "Well, the thing to do, probably, is to eat something, then you'll feel better." He flipped a can of tomatoes over to Bill that floated left to right before my eyes.

Bill opened the can, ate a few tomatoes, then felt worse. A few moments later, a few of Bill's tomatoes reversed course and flew out and into a barf bag that minus the can floated right to left before my eyes. Oh-oh!

As we started to work, we found ourselves in a situation of absolute rigid control and micromanagement where we were not given any latitude on how we performed our work. This irrational micromanagement stunt lasted until about halfway through the mission.

Up to then, our work efficiency suffered. Unfortunately, the experimenters and NASA lost data that would otherwise have been obtained. We could not understand it because the controllers that we personally knew on the ground were friendly, rational, intelligent people.

Yet we continued…

When we finally entered Skylab, we discovered that we had company. Three figures were wearing our space suits, complete with our Skylab mission emblems and name tags, and were already engaged in normal Skylab activities. They were exercising on the bicycle ergometer, undergoing a medical test in the lower body negative pressure device, and performing a normal human bodily function in the Skylab lavatory. Before leaving, Al Bean and his crew had some fun setting up this practical joke.

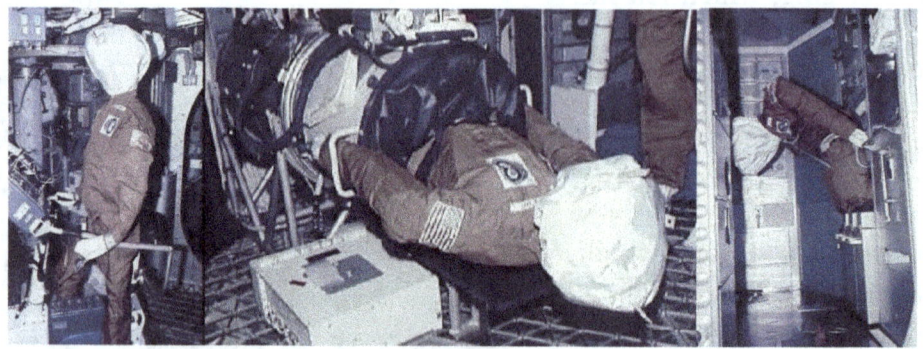

This crew greeted us and showed us how it was to be done.

But, after a few days, I lost about three pounds of water and felt good again. We really appreciated their humor and good wishes behind it, but our schedule was tight, so we initially just left our alter egos in place and got to work. When we saw one of them out of the corner of an eye, it seemed like we had some additional crew up there. Unfortunately, not one of them did a lick of work.

Even though we were thrust into a whole new environment, Skylab felt like a home away from home. However, when I looked in the mirror, a pumpkin looked back, a round redhead with bright red eyeballs. No longer countered by gravity, my heart continued to ram blood through my arteries up towards my face. It felt like I was lying on Earth with my feet higher than my head.

Whenever we were within radio range of a ground station, we were greeted by a continuous stream of detailed, rapid-fire instructions. Because of our previous thorough training, a good part of this micromanagement was unnecessary, frustrating, and only put us farther behind. Many weeks later, one of my crewmates expressed his frustration, anger, and pain with clenched fists, rigid body, and bulging teared eyes. "Is THIS what I've worked so hard my whole life to achieve?!" The other two of us felt the same.

Clearly, this situation was not constructive for anyone. However, through great effort, our work pace was accelerating. We were determined to sprint this micromanagement marathon.

In our spacewalk repair outside of Skylab, we had to cut away many layers of Mylar that were thinly coated by aluminum to get to an instrument itself. Many hundred pieces of this material were blown away by the gas escaping from our suits and quickly tumbled toward the horizon. When they reflected the red light from the setting sun, it appeared that twinkling red lights were following us, and we commented on it. Of course, a few tabloids made mention of the aliens that were in hot pursuit of our space station, a manufactured story that gave us a laugh.

A spacewalk itself is enjoyable. Head down, your glide over Earth at a very serene five-miles-a-second. The laws of Sir Isaac Newton give you complete "intellectual confidence" that you're up there to stay.

But, when you move away from the spacecraft, look straight down at Earth 270 miles below and feel or see nothing around or immediately around or below you, a little voice comes out of nowhere, "Hey you, suppose that Newton guy was just a little bit wrong?" Each time I'd smile, then get back to work.

"Just love a spacewalk!

Comet Kohoutek

Visual observations of Comet Kohoutek were made and recorded by our drawings in flight and post-landing. A sunward spike appeared almost immediately after the perihelion (closest approach to the Sun). Jerry and I first observed on a spacewalk on December 29, 1973, one day after perihelion. We did not anticipate it even though we had received an email from the ground after launch that the comet's original tail would "fan out" as it went around the sun. Some of the larger and heavier material in the comet's tail was slower to swing around the sun and formed the sunward spike that we observed.

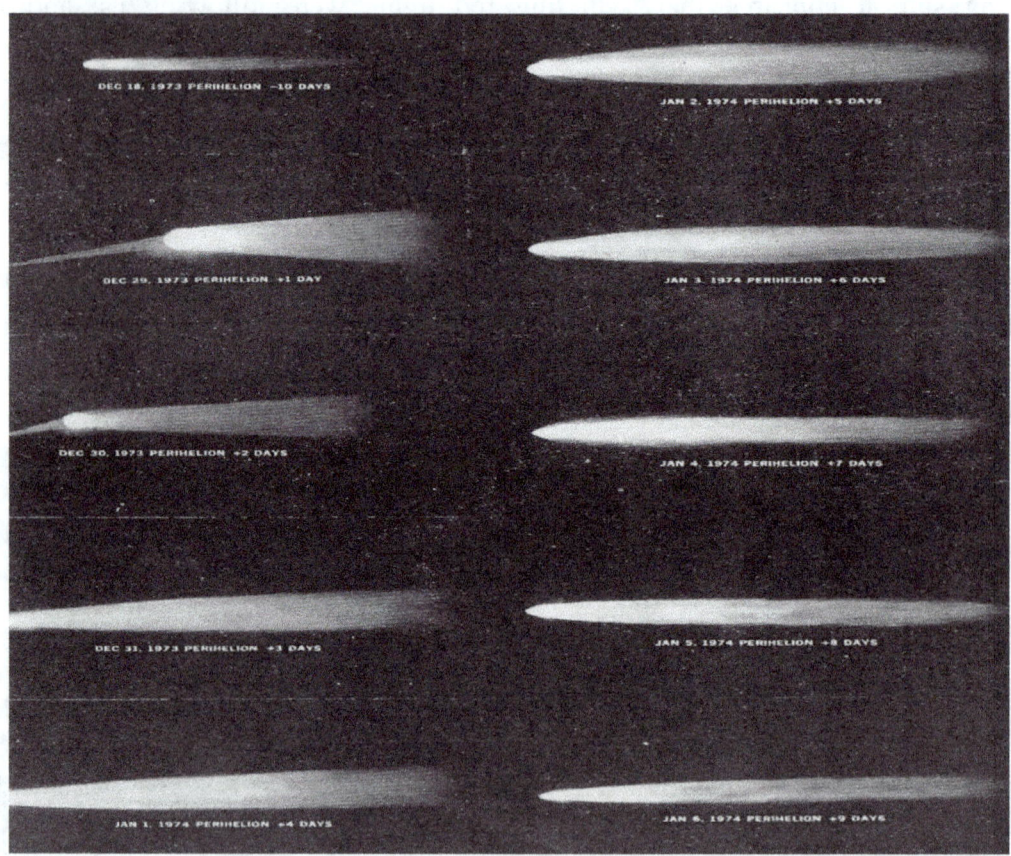

A short history of Comet Kohoutek as we saw it then drew it.

Space, Ever Farther, Ever Faster – Now!

Comet Kohoutek seen through the eyes of an astronomical telescope.

Ed Gibson

Our Skylab III Mission Patch

Our crew patch, which Bill designed, shows Skylab's three primary mission objectives: medical experiments to understand the effects on humans of long-term space flight, the study of Earth's resources represented by the tree, the study of science, especially the Sun, represented by the Hydrogen atom, which is the Sun's principal constituent and. The rainbow represents the biblical reconciliation by man between our resources and our science. (And, if you are a fan of Mickey, rotate it a quarter turn left.)

MEDICAL OBSERATIONS

As we `trained for our medical experiments, we were provided a solid understanding of their objectives and came to appreciate the skill and dedication of the experimenters. They were highly cooperative and supportive during our training and the flight itself.

Skylab was a significant challenge for the medical community as they tried to answer the question, "What happens to the human body in long-duration space flight.

Mineral Balance

First, what happens to some of the essential elements in our body: calcium, magnesium, nitrogen, sodium, potassium, and phosphorous? Accurate daily measurement of these minerals that we consumed and then excreted in our urine and feces was required, which was accomplished using returned daily samples at the end of our mission.

Complicating the food system was the extension of our flight from 56 to 84 days, an extension that was not supported by the amount of food initially launched on the space station. Also, we had no space in our CM to carry up all the additional food required.

Enter the *"small food bar."* Every third day we ate food bars that contained all the minerals and calories required and needed only a minute to consume. To make our food semi-enjoyable, we also had items that did not contain any of the above minerals, had some calories, and were enjoyable to eat. The three of us became addicted to butter cookies, which were in short supply and soon became the Skylab monetary unit.

The food system on Skylab was excellent, which we enjoyed two out of every three days. On non-food-bar days, we had provisions for room temperature and hot foods including steak and lobster Newburg. We also had frozen desserts like ice cream. All our drinks were rehydrated and contained in squeeze containers, as shown at the top of the tray (below).

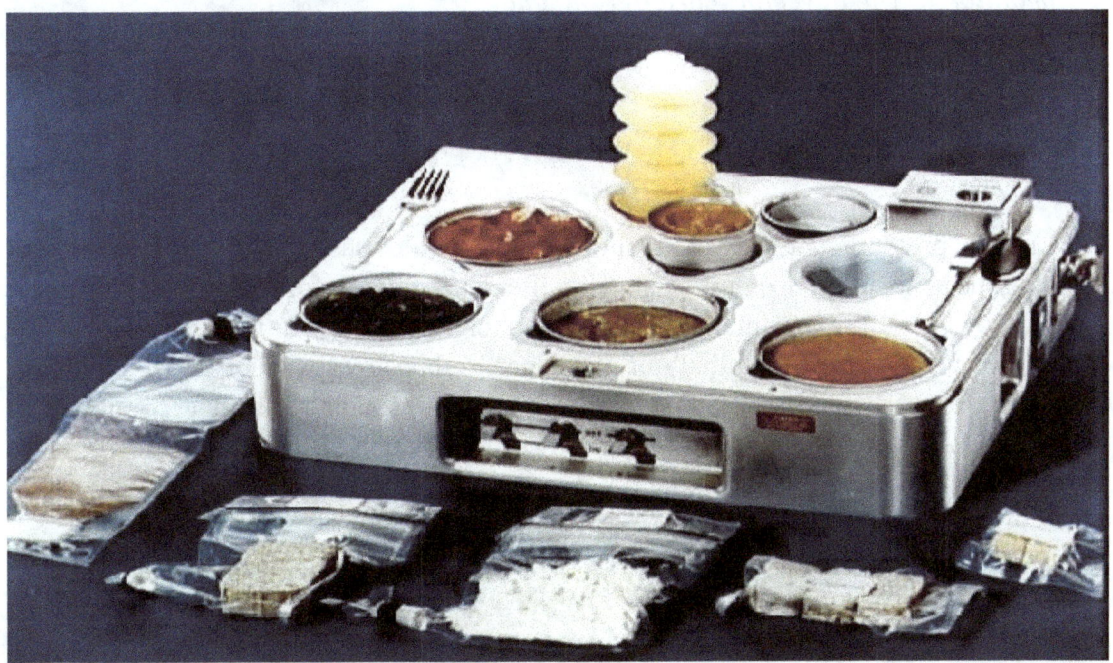

Like the previous two Skylab crews, the results showed that we each lost calcium at a steady rate (about half a percent of total body calcium per month), which resulted because we did not have gravity to stress our bones, and they adjusted accordingly. We also lost nitrogen and phosphorous, which showed that we also lost muscle mass even though we exercised more than the previous two crews. These results are somewhat consistent with what is experienced by people confined to bed rest on Earth, but the loss rates are not necessarily the same. However, other than these losses, our mission clearly had no resemblance to bed rest!

Fluid Shift

As mentioned previously, upon entering zero gravity, the upper part of our bodies became engorged with blood as one would experience down here with their feet higher than their head for several minutes or longer. It was uncomfortable at first, but we soon mentally adapted to it. However, we noticed that we always felt better throughout the flight for one or two hours after exercising on a bicycle ergometer, which drew blood down into our legs.

Lower Body Negative Pressure (LBNP) Experiment

With a seal around our waists, the air pressure around our legs was lowered in stages, and our heart activity was monitored via a vector cardiogram and blood pressure cuff.

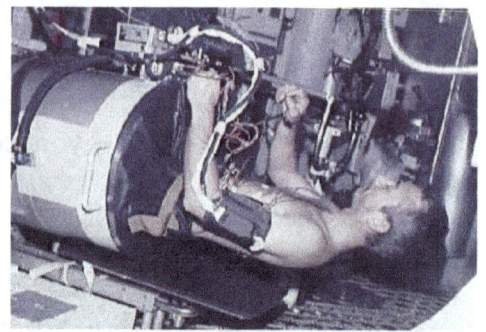

LBNP Experiment in use by Owen Garriott on Skylab II

If the pressure was lowered enough, we would pass out (syncope). It was analogous to standing up again in gravity after laying down for a prolonged length of time. However, the experiment was always concluded on time or if severe presyncope conditions arose. We found that our resistance to this stress was lowest at about 4 to 6 weeks into the mission, after which our cardiovascular resistance continued to improve to the end of the mission. The lowering of this resistance did not influence our ability to perform our work.

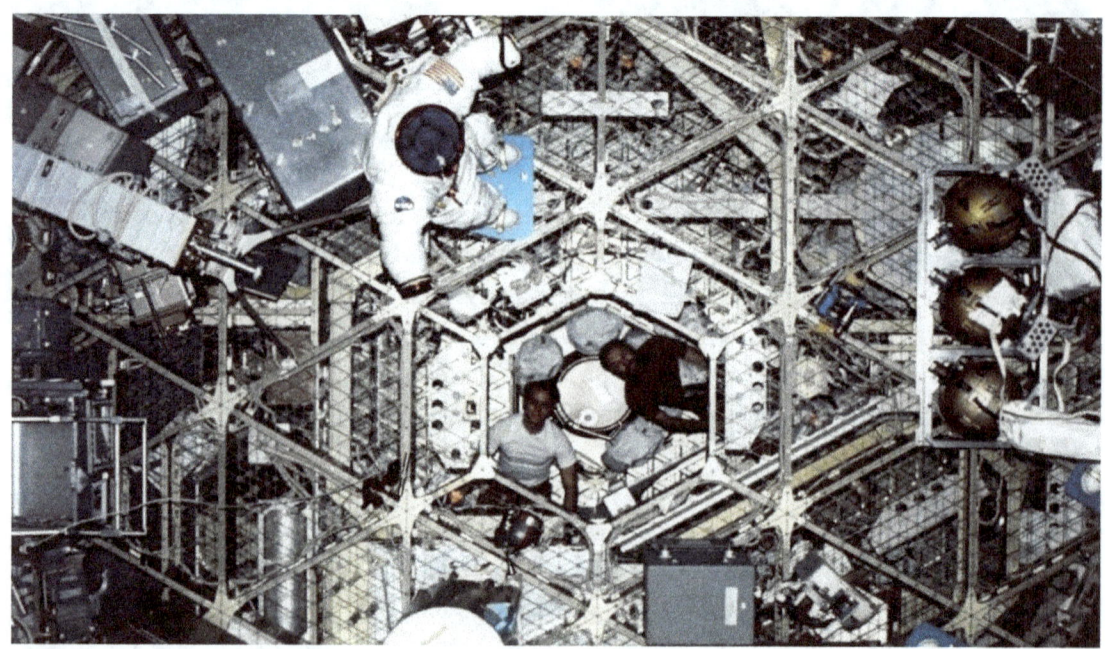

Ed and Jerry peering up into our 22-foot diameter upper section of Skylab

At Skylab's conclusion, we recognized that more work on space sickness is required to define future crewperson's susceptibility to it, effective preventive measures, and medications to relieve it. It is my non-medical opinion, however, that one will adapt to continuous life without gravity in time, some people more quickly than others.

Vestibular Effects

Although I could keep stomach awareness below a distractive threshold by moving carefully in the first few days, eventually, its threat disappeared. I did experience a sensation of tumbling in the rotating chair, but it never led to stomach awareness. Neither did acrobatics in the large workshop where I did 15 to 20 forward or back flips at a time. I did experience severe nystagmus, which was flickering of the eyes as they tried to keep up with the fluid racing through my ears semicircular canals, but it never induced any stomach awareness.

When moving around inside Skylab, I never attached any direction to my motion other than across. But, after looking out a window in the direction of the ground for several minutes and turning my gaze back inside, I experienced the sensation of looking down inside Skylab and grabbed a handhold to keep from "falling." It occasionally provided some self-amusement.

I did notice a "ballistocardiograph" effect a couple of times when I was trying to take pictures through a window and was just holding on to an adjoining structure with fingertips only. It felt like the whole Skylab cluster had a lightly beating heart at around 60 beats per minute.

Because of our high orbital inclination (50°) and height, 270 miles, we just grazed the bottom of the inner van Allen radiation belt many times over the course of our 84-day mission. Thus, we each experienced a higher cumulative dose of space radiation than any previous American or Russian crews (about 15 REM).

Like astronauts on missions to the Moon and back, we experienced light flashes originating from high-energy particles penetrating our eyes. They were well correlated with the South Atlantic Anomaly, a region where the magnetic field dips down lower than other locations and charged particles follow it.

When a major flare occurred on the Sun one night, I saw flashes that looked like many pollywogs; concise ones, many of them of low intensity. Also, I saw one large green flash, not a slightly green flash but a good old St. Patrick's Day bright green flash.

We slept in cots, each outfitted with four straps that held us down and gave us the illusion that we were sleeping in a bed. On several occasions, I tried to sleep while floating free in the large volume of the station. However, as I floated like I would in water with my arms out and knees bent, eventually I would slowly mash into a wall and realize that it would soon happen again. Sound sleep could not come as I waited for the next slow motion impact. However, if I had a leg or arm just lightly restrained, I could easily enter a deep sleep.

It was a surprise to us that we had no major illnesses. We were working hard most all the time and got rather tired. We stayed tired for about the first half to two-thirds of the mission. If we had done that on the ground, I don't think we would have gotten by without getting at least a bad cold.

EARTH OBSERVATIONS

It was enjoyable for us to use our intelligence to select data. Our Earth presented us with a continuous, ever-changing display of exciting features. A few examples are shown below.

Sometimes Jerry and Bill functioned as human computers in this highly controlled point-and-shoot dictated sequence of data gathering Earth Observations data. Data on the same site was simultaneously collected on the ground.

When we were over South Island, New Zealand, and the sun was at a shallow angle, the Alpine fault was obvious (below). It is so straight and well defined that it appeared to have been scribed using a giant ruler and plow.) It indicates the line along which earthquakes originate as one of Earth's tectonic plates slides in a halting fashion past another.

Space, Ever Farther, Ever Faster – Now!

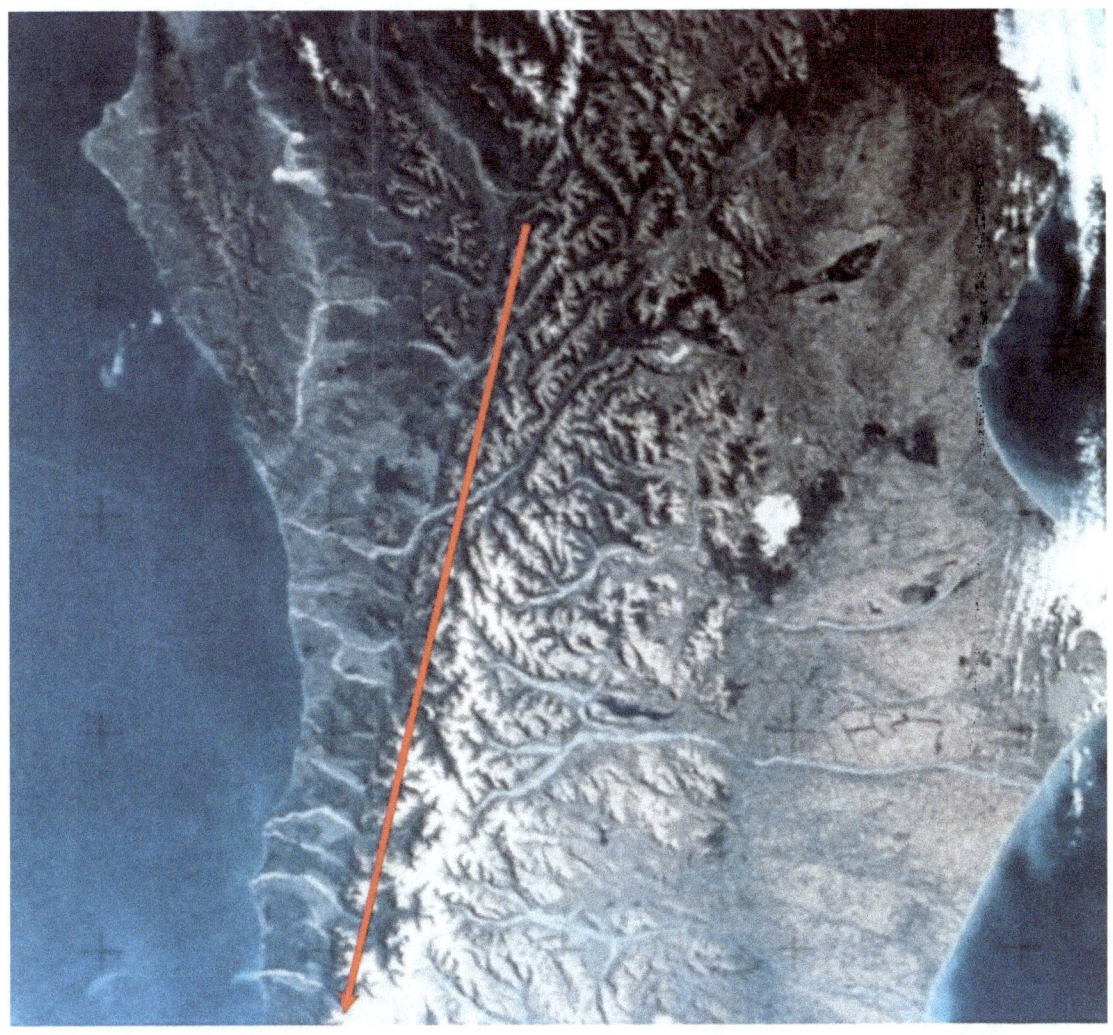

Alpine Fault, South Island, New Zealand

Over here on the west coast of America, we have the San Andreas fault (below) that is on the opposite side of the shifting Pacific tectonic plate than the Alpine fault. As with the Alpine fault, it is also evident from space.

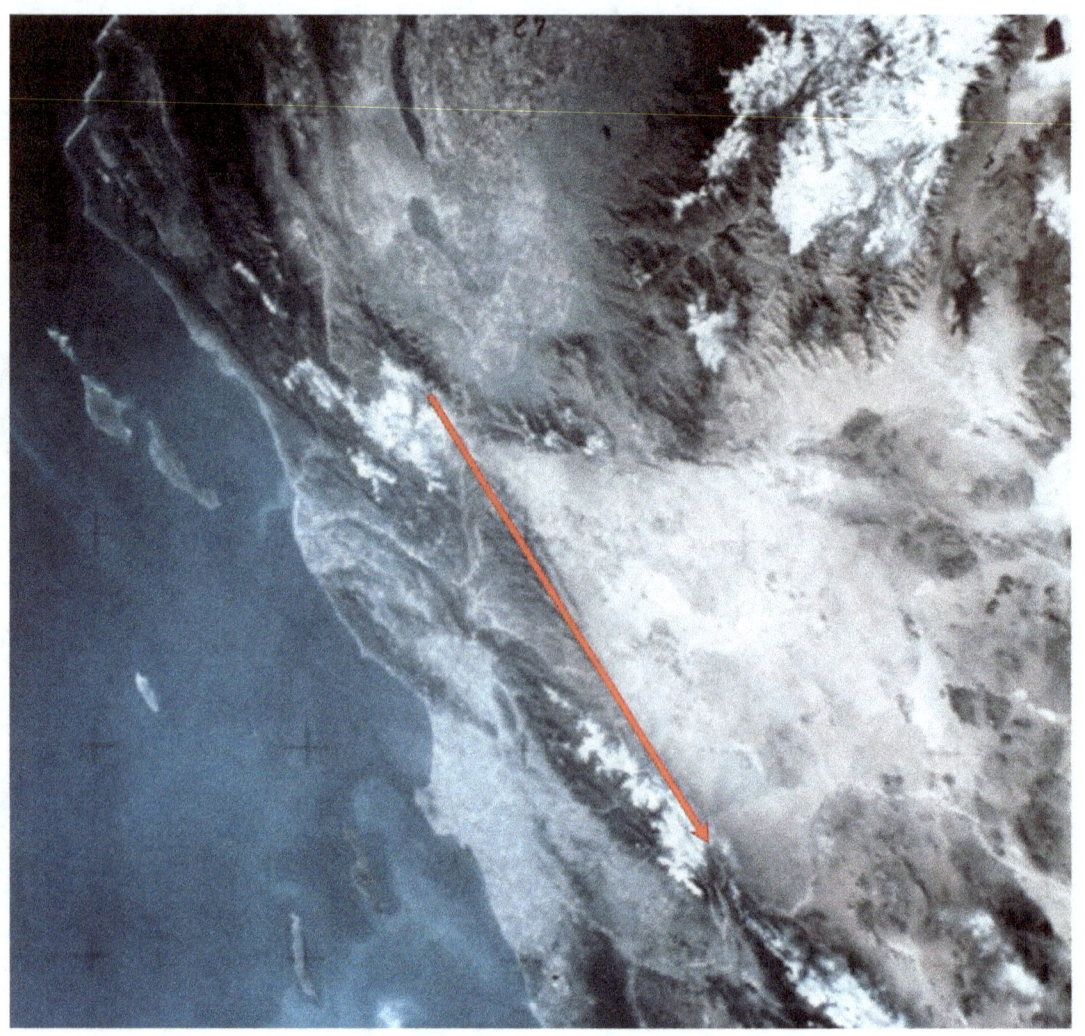

San Andreas Fault, Southern California

We observed another interesting feature in the cloud pattern over the Island of Cyprus. When a steady wind blew from the north over the mountains on the island, it created a wave pattern in the clouds like that on the ocean's surface.

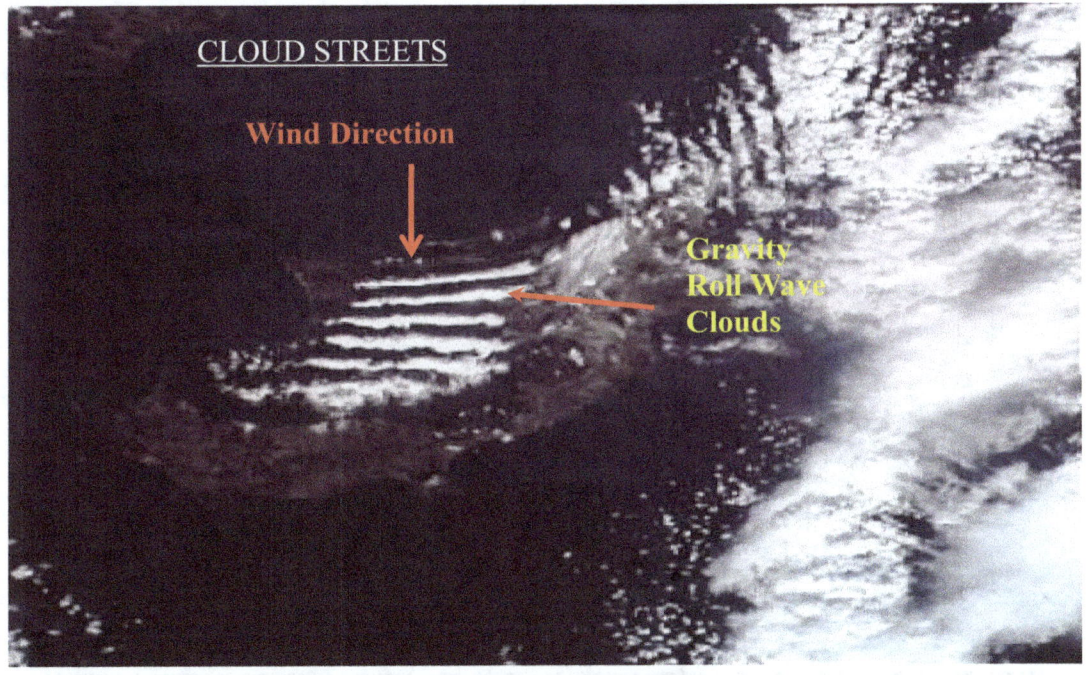

Island of Cyprus

As the air ascended into the thinner and cooler atmosphere above, moisture condensed out to form a cloud, then disappeared as the air again descended. Several such distinct, consecutive roll clouds appeared.

SOLAR OBSERVATIONS

Our solar observations were like both forms of Earth observations. We had prescribed procedures, but we could also deviate from them if we saw a situation of greater scientific value using the same instruments and their displays.

Ed at controls of Solar Telescopes

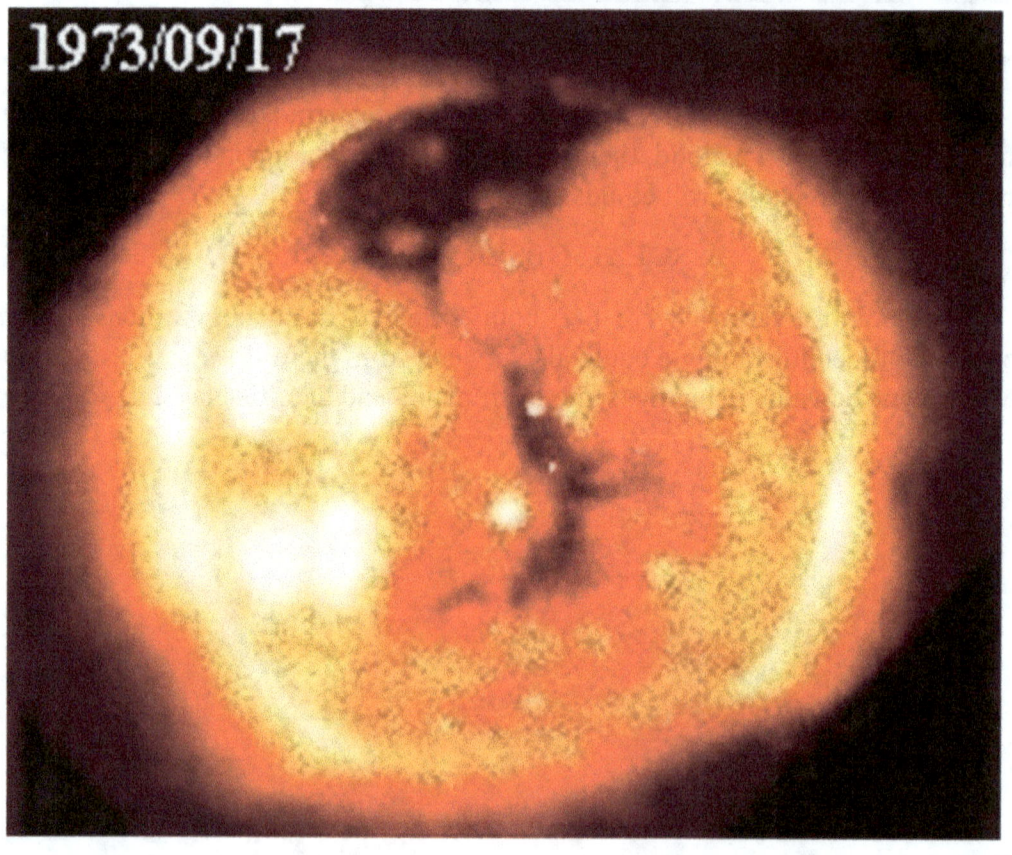

Credit NASA-MSFC - Photo recorded by the ATM

Features seen on this solar disc are:

1. High-temperature active regions that are usually associated with sunspots are on the left portion of the solar disk
2. Small bright points can be seen across the disk
3. Dark areas at the top and central portions of the disk are called coronal holes, which have a lower temperature than their surroundings and from which much of the solar wind originates (their vertical magnetic fields allow the escape of charged particles)

With so many options on what, where, and how to operate the seven high-quality instruments, considerable thought and planning were required to use our limited film wisely. It is a credit to all the scientists and engineers involved that the ATM revolutionized the field of solar observations at that time. The ATM observed and recorded data on newly discovered minor bright points seen across the sun's surface in UV and X-rays. Also observed was the liftoff of huge prominences on the sun's limb, events called Coronal Mass Ejections that threw massive amounts of material off the sun and out through its atmosphere (Corona), as well as the very early phases of solar flares that explosively released large amounts of energy into the solar atmosphere. A few examples of the images acquired are shown below and on the following pages. On features on the x-ray image of the solar disc shown on the preceding page are:

Ed Gibson

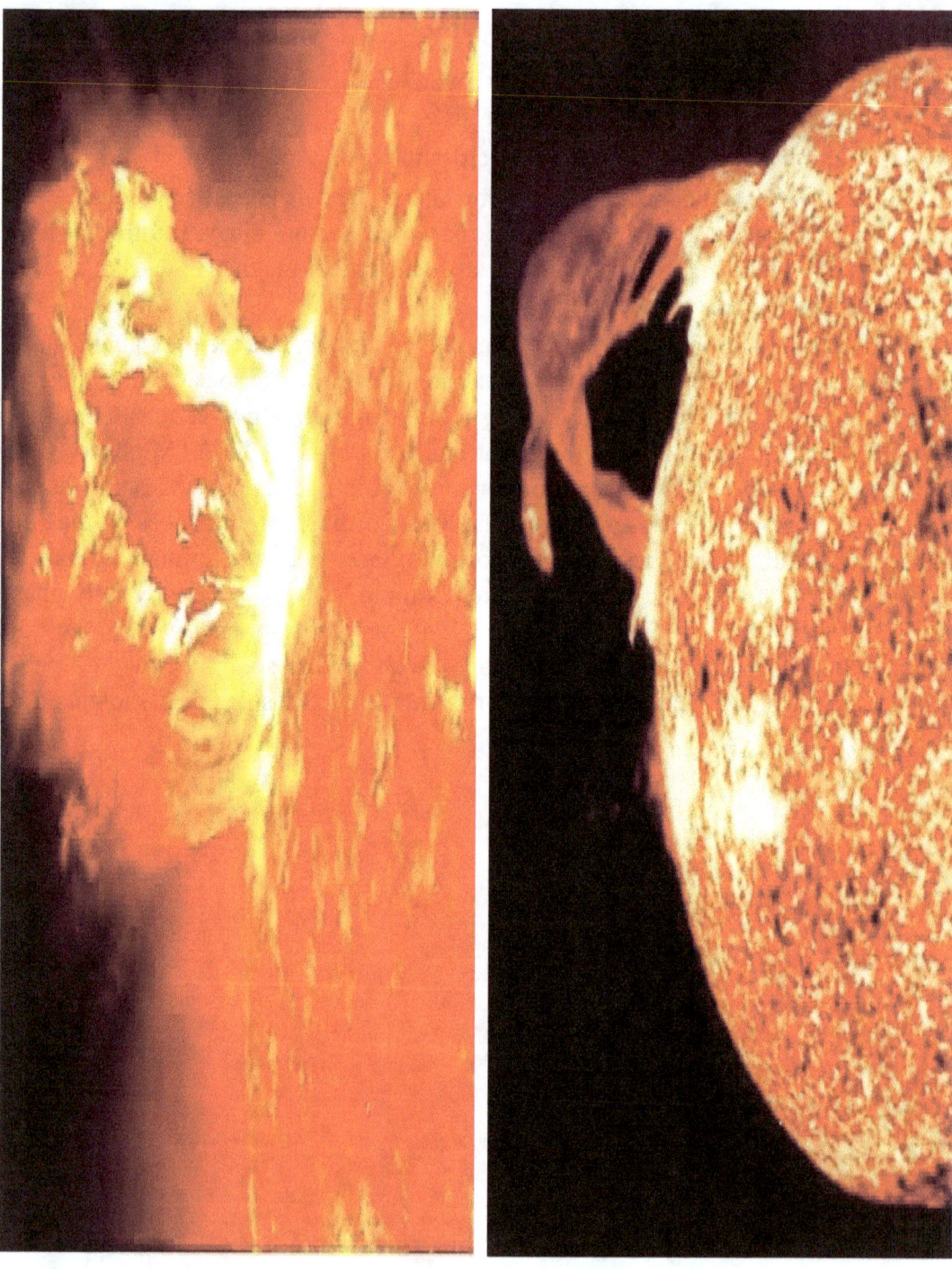

Ejection of hot ionized gas by a flare   *Solar prominence 12/19/1973*

With the sun covered by a disk (or the Moon) to block out its bright light, the faint glow of the sun's outer atmosphere, the Corona, was clearly seen (right). False color is added here to show differences in light intensity. The coronagraph filtered the light much more strongly closer to the sun so the larger geometry of its outer features could be seen and related to the ever-changing features on the sun's surface.

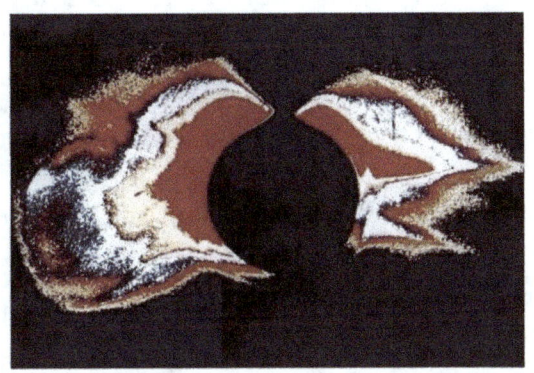

Solar eclipse seen on command from Skylab

Skylab has provided more understanding of the dynamic corona than the observations made over centuries before. When trying to observe the corona from the ground, an observer must wait for one of those rare times when the blocks out the disc of the Sun. If the observer is located on Earth at just the right spot, is above much of the atmosphere on a mountain top, which is usually remote, the weather is accommodating, and the observer has been able to reposition all his heavy, complex telescopes and recorders to that right spot, ` can then see the Corona. If all these conditions were met, the observer might be rewarded with a minute or two of observing time.

On Skylab, we could observe the corona anytime we were in sunlight. One result of these observations is realizing that the sun's corona is sometimes highly dynamic. Several experts in the field highly doubted this dynamic activity existed before Skylab data was returned. After all, when they could make observations for only a minute or so every 5 or 10 years, they had no way to speculate on changes that covered just seconds, minutes, or even hours.

Attempts to record the very early phase of a solar flare, much time was spent observing "active regions" in which these explosive events occur. As I monitored one active region on January 21, 1974, the frequency and intensity of 'mini-flares" began to increase.

It reminded me of a pot of water beginning to boil. I glued my vision on it and was rewarded when a significant flare began to erupt, which gave me the confidence to shift instruments into their rapid data-taking modes. This decision could not be made lightly since all, but one device, recorded data on limited amounts of photographic film. We did record the birth of the flare then, but the spatial resolution of instrument's image was insufficient to provide the level of detail we sought to fully understand the flare's birth when the data was analyzed after the mission.

Our sun is not as remote as it usually appears. Energetic particles can explode from the solar surface, rip through its atmosphere, enter and dominate the solar wind, and sometimes, even impact us here on Earth.

ATM has given us a much greater understanding of the processes involved in these explosions, methods to predict their occurrence, and ways to mitigate the related adverse impacts on our lives.

ATM also observed and recorded data on newly discovered minor bright points seen across the sun's surface in UV and X-rays. In addition, ATM observed the liftoff of huge prominences on the sun's limb, events called Coronal Mass Ejections, which threw massive amounts of material off the sun and out through its atmosphere that we in observe the solar wind on rare occasions. Skylab has provided more understanding of the dynamic corona than the observations made centuries before.

At mission conclusion, Skylab III had recorded 75,000 new telescopic images of the Sun in X-ray, ultraviolet and visible light. The whole NASA team and those in the realm of the experiments (scientists and supporting personnel) deserve, even 50 years later, the gratitude and recognition of today's scientific community for their contributions to the field of solar physics and its many beneficiaries.

At the end of our solar telescope's work, it gave us a wink and a smile!

It's Christmas!

One day around Christmas, Jerry disappeared for a few hours. When he reappeared, he proudly held up the spindly contraption shown to the right.

"Look guys, we have a Christmas tree!"

What is Christmas without a tree? We could not grow one, so Jerry made one. Our "tree" was made from equipment used to hold together food cans. The star on the top was a cutout representing comet Kohoutek.

Our Christmas Tree

It's the best he could do with limited time, budget, and store availability.

But what would our Christmas have been without the initiative of Jerry!

A Surprise

When the smoke cleared at the end of the mission, it was calculated that we had averaged slightly more work per day across the whole mission than the second crew, who set the standard by which we were being judged. This was unexpected because, it was not until halfway through the mission that Bill came up to his usual high efficiency and we were freed of the micromanagement stunt.

What a great surprise!

The dedicated Mission Control Teams that worked hard to provide excellent support!

Re-entry

Our first taste of "gravity" occurred when we made our first rocket burn to decrease our speed for entry at the end of our mission. The acceleration created a sensation of tumbling but in no one direction. We splashed down into a calm sea with no wind. However, we ended up in what NASA called Stable 2. That is, hanging upside-down in shoulder straps, just as we had experienced in training.

Landing

Stable 2

Space, Ever Farther, Ever Faster – Now!

Once back on the deck of the Aircraft Carrier USS New Orleans, we felt depressed. No matter how hard we pushed off, we could no longer float. And no matter where we went, we were aware that we once again had to haul along massive amounts of meat and bone. However, unlike some crews after long durations in space, we could walk as soon as we landed and suffered no immediate debilitating effects.

However, we cheated. We wore G-suits that we inflated before getting off our backs and leaving the spacecraft at the medical team's instructions. These suits compressed our legs and kept the blood from leaving our upper bodies. We clearly understood that stepping out of the Command Module and immediately crumpling to the carrier deck was not considered good public relations.

Once back on the ground, movements were made with caution. Rolling over in bed, moving an arm or leg, and walking had to be anticipated and not done impulsively. Our unsteadiness dissipated after a few days and was completely gone after two weeks as measured by our performance on a balance beam.

Summary

On Skylab III, we completed 1,214 Earth orbits and four EVAs totaling 22 hours, 13 minutes. We traveled 34.5 million miles (55,500,000 km) in 84 days, (officially listed as 2017 hours and 16 minutes of *"high-performance rocket time."*) We slept through about 30% of this rocket time.

Our 84 days in space set a world space endurance record, which, unfortunately, lasted for 21 years as an American record. We wanted and expected this American record would be broken long before. But when it was finally broken, it was by a competent and dedicated America, Norm Thagard, but he had to do it on a Russian space station.

The backup Skylab now resides in the Air and Space Museum in Washington rather than in orbit, where it could have possibly served as a nucleus for our next station at a much earlier time and lower cost. The remains of the Skylab space station itself are now spread across the largely uninhabited grasslands of south-western Australia and in the adjoining Indian Ocean, where it eventually impacted on July 11, 1979. The Australians have been very gracious about our scattering the remains of Skylab across their country, and they didn't even bill us. In fact, they have had celebrations in honor of the occurrence. Please see References (31-33) on Skylab.

Although unrecognized at the time, Skylab accomplished the painful transition from Apollo to Space Station operations. Skylab I accomplished excellent "save-the-station" repair work, and experiments had to take second priority. Skylab II focused on operations and eventually reached a high efficiency that kept Mission Control scrambling to send up more work. Skylab III was expected to reach an efficiency early in the mission that was not reached until much later in the two previous missions. We paid a harsh penalty for those incorrect expectations and the micromanagement but, in the end, we worked with Mission Control to achieve an average work output per day over the entire mission that was slightly larger than that of Skylab II. We also established the more efficient mode of operations that benefitted future missions.

In the larger picture, we Americans should feel pride that we made, often with great difficulty, the positive and indispensable differences in our technical, geopolitical, operational, and scientific areas that led to outstanding successes in the Mercury, Gemini, Apollo, Skylab, Apollo-Soyuz, and International Space Station Programs.

Space, Ever Farther, Ever Faster – Now!

Skylab as we last ssaw it at Mission's end

Ed receiving the NASA Distinguished Service Medal from President Nixon

Our Long-Term Future

In the very long term, which is beyond the scope of this book, what is in store for the US and all humanity in future space missions and major developments?

In general, it is desired that we will:

1. Focus on international cooperation to reduce global tensions.
2. Reduce costs to make space flight more financially acceptable and utilized.
3. Explore farther and farther away from Earth (moon, solar system, and farther out to exoplanets)
4. Utilize new territories already explored.
5. Repeat 1 thru 4 on ever larger scales as we truly become a space borne civilization.

Eventually, Human will make all the above steps. But will we, the Human of today, take the next steps forward or just keep our eyes focused on our shoes and leave everything up to the Human of the distant future?

Fortunately, today, there are some positive initiatives being taken.

Progress or not? It's our choice.

Ed Gibson

Human is stronger when all factions cooperate.

Chapter 13: Apollo Soyuz - The first Joint Mission with the Soviets

Within the Apollo Soyuz Program, respect and friendliness existed throughout. It was hoped, and many believed, that this evident goodwill and close cooperation would extend upward into the overlying political structures. But as of today, it has not!

Crew of Apollo Soyuz

Front row: Deke Slayton, Vance Brand and Valery Kubasov

Demonstrating that our two dissimilar spacecraft could dock in orbit was relatively straightforward. The Docking Module between the two spacecraft was designed together by America and the Soviet Union and built in America. In addition to enabling docking, it also served as an airlock between the 100% Oxygen at 1/3 atmospheric pressure in Apollo and the Oxygen/nitrogen combination at 2/3 atmospheric pressure in Soyuz.

The human side of the mission would be more challenging. Training for the mission together and living in each other's countries helped break down cultural barriers. People in the same line of work dedicated to the same objectives learn to appreciate one another for their common skills, ambition, and humanity.

The differences in the two languages were a challenge not easily overcome. Vance Brand was my immediate next-door neighbor in Nassau Bay, Texas, and a close friend. I felt a great deal of empathy watching him trying to become reasonably proficient in Russian. Nope, it wasn't easy for the crew members on either side. However, each side made a hard and honest effort to learn the other's language and adopted the practice of constantly communicating in the other's language. This practice was the key to a much-improved information flow and mutual understanding.

NASA Flight Directors and the Program Manager were each assigned a translator on the ground. During joint mission simulations, Flight Directors spoke in their own language, then it was translated. It worked well, and each side became so comfortable with the translators that they relied almost exclusively on their work in the mission.

However, outside political interests continued to be a disruptive force. But minds have been opened, and trust has primarily replaced suspicion. This partnership has grown to include Canada, Europe, Japan, and many other nations, for example, on the International Space Station. Future human expansion from Earth into our solar system will most likely come about through international efforts.

June 2023 - Now, more than ever, the *Apollo Soyuz Cooperative Spirit* is essential for our peaceful coexistence. Leaders with dedication to strong cooperation are sorely needed for the sake of world peace. We need to again rebuild it!

It was and continues to be the hope and objective of most of those who contributed to the success of Apollo-Soyuz that America and Russia can now come together to reestablish our continued peaceful coexistence. Certainly today, most of the astronauts and cosmonauts and the respective Mission Control Center personnel who were at the core of our spirit of cooperation and teamwork are now retired. However, we cooperated then, it worked, and we make it happen again,

Of course, we must overcome the drive to make the now independent state of Ukraine once again fall under the total domination of the Soviet Union. Killing one another on the battlefield is not a solution. Entering into an honest, open discussion to understand the human drives behind our opposing military demands can lead to an acceptable coexistence without the massive death of war if, this outcome is really desired.

After the mission's inflight activities concluded, Soyuz undocked and re-entered while Apollo remained in orbit for three days and performed science. The ultraviolet detector carried on Apollo recorded extreme ultraviolet sources outside the solar system, including the hottest white dwarf known, which was a first and a surprise to many scientists. They also recorded data on the first pulsar ever observed outside our galaxy, almost ten times as bright as any pulsar observed to that time.

In addition, on their mission they completed nearly all their 110 earth observations, which were made simultaneously by six groups of scientists on the ground, on ships at sea, and in aircraft. They observed ocean currents, ocean pollution, iceberg movements, desert geography, volcanoes, shoreline erosion, and vegetation patterns.

The crew in their Apollo CM re-entered on July 24 and experienced a problem when the CM's apex cover and drogue chutes failed to jettison automatically and had to be accomplished manually. Because of a spacecraft configuration issue, some tetroxide from the reaction control jets on the CM entered the spacecraft and caused considerable crew discomfort. Also, after splash-down, they ended up inverted in the water; that is, Stable 2. (Maybe they were just jealous of our enjoyable experience on the previous mission).

This mission was gratifying for Deke. As mentioned previously, he was one of the "Original Seven" Mercury astronauts selected in 1959 but never cleared for flight because of a perceived heart condition. As soon as he was finally approved for flight, he started training for Apollo-Soyuz. Everyone in the Program was happy to see him fly, and he certainly enjoyed the experience. In all those years before he was cleared for flight, Deke was driven to contribute wherever he could. Deke was not a quitter!

Soon he was appointed the Director of Flight Operations, where he was in charge of Mission Control, Aircraft Operations, and the Astronauts. Immediately, he became responsible for many personnel including about 40 head-strong astronauts, each one of which was always ready to charge off in their own directions. But he tightly reined in every one of us because he was tough and very mission focused.

He gave us his full support and encouragement if you were there to advance the mission. But if you were there more to promote yourself or some other cause, he'd rip out the thrower and turn you to a crisp in nothing flat. He was the right guy for the job! I regard Deke as one of the most outstanding leaders I've ever personally encountered. As he demanded an overwhelming focus on the mission, he was tough but fair, harsh but kind, someone you respected, trusted, liked, and feared all at the same time. Working for Deke was a privilege and an exciting education. Also, Tom

Space, Ever Farther, Ever Faster – Now!

Stafford, who served as Deke's assistant, was like Deke and always performed with strength, competence, good intentions, and consistency.

Credit: Wikipedia: An *artist's drawing of Apollo spacecraft, its joint Docking Module and the Soyuz Spacecraft*

Chapter 14
The Space Shuttle

After the premature termination of America's lunar explorations driven by decreased NASA funding, we set out to build the reusable Space Shuttle that promised a lower cost of space operations, which was clearly needed to enable future giant strides into space, not just one single modest step at a time. The Space Shuttle development offered a tremendous challenge to the Aerospace workers who remained from the days of Apollo and the relatively new workers who came on board with great enthusiasm and technical skills but were acquiring the necessary space flight and operational skills. Again, success required ample Vision, Courage, and Commitment.

Upon reflection, our Space Shuttle was a tremendous technology development and contributed greatly to our sustained progress in space. It was NASA's workhorse for 30 years from 1981 to 2011. It flew 135 missions to construct the International Space Station (ISS), launch many satellites, launch the Hubble Space Telescope (HST) and then service it on five additional missions. This telescope has revolutionized the field of astronomy.

TDRS-A Liftoff

The start of the maiden orbital flight of Space Shuttle Challenger occurred on April 4, 1983, as it launched the Tracking and Data Relay Satellite (TDRS-A). This satellite would revolutionize low-Earth-orbit communications from NASA's Dryden Flight Research Center at Edwards Air Force Base.

Space, Ever Farther, Ever Faster – Now!

Approach and Landing Tests

The Space Shuttle program started with five Approach and Landing tests made by the Shuttle Enterprise at NASA's Edwards Air Base in the second half of 1977. Practical piloting skills were required to fly the Shuttle manually. It was a "lifting body," which essentially is a glider with a low lift-to-drag ratio because of its relatively small wings. Precision was required to deorbit and arrive at the threshold of the desired landing field with just the right amount of energy remaining each time. Second tries were entirely out of the question.

In fact, every Shuttle descended at steep angles (18 to 20 degrees) that were decreased to 1.5 degrees just before touchdown. Each and every astronaut who sat in either one of the two front two seats was highly skilled and able to fly it manually with excellence. In the history of the Shuttle, the piloting results were always exceptional.

Space Shuttle piggy-back ride on 747 aircraft on the return from Edwards back to KSC

John Young and Bob Crippen were the first to fly the Shuttle in orbital flight, which was the only orbital spacecraft ever to be launched with a crew onboard for its initial flight. The first five flights were made in the Columbia Orbiter, which included the first use of a robotic arm in the payload bay from Canada appropriately called the "Canadarm," the first DoD dedicated mission, and several Comsat deployments. Columbia also flew the first Spacelab payload bay experiment module on STS-9.

Subsequently, Challenger, Discovery, and Atlantis joined the fleet that flew the next 19 missions. These included the first Shuttle spacewalk (Story Musgrave and Don Peterson), our first American woman in space (Sally Ride), multiple Comsat deployments, our first African American astronaut in space (Guion Bluford), and the first night landing.

The Rogers Commission, established by President Reagan, found that both NASA's safety decision-making and the culture in which it was embedded were flawed and major contributing factors. Critical information was adequately defined, but the deciding managers did not take appropriate actions.

The Commission stated that NASA had violated its own safety regulations. On the day of launch, they failed to adequately consider the low-temperature restrictions on the SRBs and ignored the impassioned warnings of knowledgeable engineers. In addition, the problem with the SRB O-rings, which was detected very early in the program, was not aggressively addressed by Morton Thiokol or NASA even though the impact of failure could be catastrophic and was in the highest level of safety classification.

These technical and managerial considerations should not obscure the personal tragedies of each life lost in this and other accidents. Being on their selection board, I got to know four of the seven crewpersons launched on Challenger personally.

During the selection of Ron McNair, I had the opportunity to become very familiar with his background. As a black youth, he grew up in poverty, was skilled in football, and became interested in science. While earning a bachelor's degree from North Carolina Agricultural and Technical State University, he was given the opportunity to work at MIT in the summers and was later accepted there for graduate school, where he earned a Ph.D. Along the way, he also earned a fifth-degree belt in karate, which he then taught other youth growing up in the same situation that he had. It helped him pass along the sport's discipline that would benefit them throughout their lives. Each time he moved into a new environment, he found a church that had a basement that he could use to gather local youth and teach them the sport. He got many young men moving in constructive directions. He was a very admirable underprivileged youth who made all the right decisions!

Each of those lost were admirable people with their own compelling backgrounds, and their loss should never be treated only as a statistic.

> **Good intentioned but distracted managers put secondary issues before CREW SAFETY at a tremendous price.**
>
> *Have we FINALLY learned our lesson?*

We Continue... On September 29, 1988, after the investigations ceased and resulting action items were completed, Discovery lifted off to deploy a TDS. Subsequent missions included deployments of DoD payloads, another TDRS, the Magellan probe to radar map the surface of Venus, and an investigation headed for Jupiter.

Hubble Space Telescope (HST)

On April 24, 1990, Space Shuttle Discovery lifted off to deploy the Hubble Space Telescope. All went as planned, and scientists and the public impatiently waited for the new spectacular revealing pictures of our universe that Hubble promised. About a month later, images were released that, although better than those from ground-based telescopes, fell far short of what was expected, and the telescope was designed to deliver. How could something like this have happened with major efforts by the best technicians in the country and several billion spent? This telescope should have been perfect!

After much investigation, the problem turned out to be an undocumented little 3-mm wide washer that a technician used to check the shape of the mirror during its fabrication years earlier. End-to-end testing of the telescope before launch, which would have identified this flaw, was not done!

Finally, on December 2, 1993, STS 61:

Endeavour lifted off on a mission to repair Hubble three years and seven months after the discovery of the telescope' potentially fatal imperfection. Onboard was a revised Wide Field Planetary Camera 2 having new mirrors that would correct the light coming from Hubble's defective primary mirror.

The crew working inside of HST

Spacewalk repair at 353 miles altitude. Story Musgrave on the robotic arm and Jeff Hoffman in payload bay worked on the fifth and last repair of the HST.

Also on board was a set of other mirrors that would correct light going to four different instruments, mounted on a device called the Corrective Optics Space Telescope Axial Replacement (COSTAR), which took the position of the High-Speed Photometer in the HST, which no longer was onboard.

Ed Gibson

The spacewalks to make the required corrections to Hubble optics were difficult, complex, and lengthy. Two crews performed excellent work: the team of Story Musgrave and Jeff Hoffman performed three spacewalks, while the second team of Kathryn Thornton and Tom Ackers performed two. Over 35 hours were spent performing these repairs. Their work and that of the ground personnel that designed the repairs presents a solid testament to the exceptional work that the NASA Team was and can accomplish.

Space, Ever Farther, Ever Faster – Now!

The positive results from the repair are evident in the viewing of a single star:

From the Ground   *Before HST R*   *After HST Repair*

The Tip of the Spear, the Crew of STS 6

Seated: Ken Bowersox, Kathryn Thornton, Story Musgrave, Claude Nicollier; Standing: Richard Covey, Jeff Hoffman, and Tom Akers

A few of Hubble's accomplishments include a far better understanding of the rate at which our universe is expanding, the age of the universe, confirmation that discs of gas and dust called nebulae are the birthplace of planetary systems, black holes exist at the center of every galaxy and the far superior images that were obtained of supernovas and galaxies.

"Indubitably, ubiquitous yet specific observables to further, cogitate scrutinize, and ponder. Yes, definitely indubitable!"

Over the next 15.5 years, four more Hubble repair and upgrading missions were completed, but none was more difficult or critical then this first.

Hubble has revolutionized the field of astronomy and is still operating today, 30 years after its launch.

STS-128 Space Shuttle Discovery Landing at Edwards AFB on September 11, 2009, after a 15-day mission to the ISS

James Webb Space Telescope (JWST)

ASA's next large space telescope in line is the James Webb Space Telescope (JWST), named after NASA's second administrator who led Apollo and spearheaded many other science programs in the early days of NASA. It is a cooperative program between NASA, the European Space Agency (ESA), and the Canadian Space Agency (CSA). The cost of this 14,000-pound telescope is about 10 billion USD. It lifted off on December 25, 2021, Christmas day!

Unlike Hubble, which makes observations primarily in the visible range of light, JWST is making its observations in the long wavelengths of visible light down to and well into the mid-infrared range of wavelengths (0.6 to 28 microns; 1 micron equals 1 meter/10^6). Relative to Hubble, this range of wavelengths will enable observation of much older objects, moving away at much higher speeds and, therefore, now much further away after the big bang, and objects that are colder fainter regardless of their history. Importantly, JWST can also observe new stars and planets forming inside cocoons of dust, preventing the escape of visible light but not infrared light.

In summary, by observing infrared wavelengths, JWST is complementing and extending what we are learning from Hubble about our galaxy, star and planet formation in our universe.

Also, unlike Hubble, the JWST was launched on a European Ariane booster. It is located at a point in the Earth-Sun system that is nearly stable (metastable), about one million miles from Earth at L2 (the second Lagrange point where gravity and centrifugal forces in the Earth-Moon system balance out to zero).

However, unlike Hubble, we are betting that this telescope works for at least five years; it is designed to operate for ten years. Unlike Hubble, it will be almost impossible to service by an astronaut because of its highly distant location.

JWST 18-section primary mirror that reflects infrared light

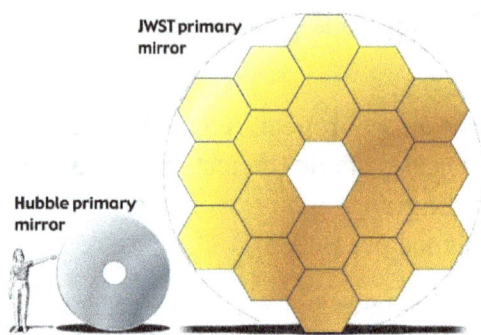

Each of these 18 sections of the primary mirror is made of a gold-coated beryllium reflector. The mirror is 21 feet in diameter and has a reflecting area (273 square feet), about six times that of Hubble's mirror. Despite this larger area, JWST weighs nearly half of Hubble.

Size of JWST primary mirror compared to that of Hubble

JWST is cooled to about -370°F to prevent it from emitting infrared radiation like it is trying to measure, which is a complex engineering feat. A sunshield is employed to block the light and heat from the Sun, Earth, and Moon. It comprises five layers of DuPont Kapton E, a polyimide coated with aluminum on both sides and doped silicon on the top two sides that face the Sun. Each of the five layers is about the thickness of a human hair.

The four instruments in JWST are (1) Near-Infrared Camara, (2) Near-Infrared Spectrograph, (3) Mid-Infrared Instrument, and (4) Fine Guidance Sensor, Near-Infrared Imager, and Slitless Spectrograph.

The JWST team is large and capable, including many NASA centers: Ames Research Center, Glenn Research Center, Goddard Space Flight Center, Jet Propulsion Laboratory, Johnson Space Center, and Marshall Spaceflight Center. Academic and Industry Partners include Ball Aerospace, Harris Corporation, Lockheed Martin, Northrop Grumman Aerospace Systems, The Space Telescope Science Institute, and the University of Arizona.

In 2018 a review of JWST found 344 single-point failures. Much work has been done, and the launch delayed reducing or eliminating as many of these threats as possible.

JWST has the potential of providing a significantly improved understanding of immense universe in which we find found ourselves.

It is now located at 932,000 miles from Earth at L2, a point of stability in our Earth-Moon system. At this distance, it will be exceeding difficult for the astronauts to make repairs if needed, which was accomplished multiple times on the Hubble Space Telescope.

As of January 7, 2023, JWST is working as planned and revolutionizing our knowledge of our universe! More such discoveries are anticipated.

SPHEREx

SPHEREx is a coming space laboratory planned to be launched on June 17, 2024. It will survey light in 96 wavelengths from 400 million stars, three-quarters of which are outside our galaxy. It seeks to understand better how our universe evolved and the commonality of life-building blocks within it.

U.S.– Russia Crew Exchanges and Mir Docking

Following the initial repair of the Hubble Space Telescope, the Space Shuttle Program struggled to satisfy its many diverse requirements, including those from SPACEHAB, Spacelab, microgravity experiments, Wake Shield Facility, SPARTAN, Space Radar Laboratory, ATLAS, numerous ISS assembly and resupply missions and early in 1995, a rendezvous with the Mir Space Station. After the Soviet Union dissolved in 1991, both Russia and the USA wished to reestablish the goodwill and cooperation in space first established on the Apollo Soyuz mission twenty years earlier.

When cosmonauts and astronauts first exchanged crews, Norm Thagard was the first American to fly on a Russian Soyuz rocket when he flew to the Mir Space Station on March 14, 1995. He remained there for 115 days, breaking the American space endurance record of 84 days set on Skylab III.

Shuttle Atlantis docked with Mir Space Station

Norm returned on shuttle Atlantis on STS-71, the first shuttle to dock with the Mir Station. Eight more such dockings followed.

Over the subsequent 43 missions, the Space Shuttle fleet continued as the work horse for space station construction and science payloads until the STS 107 mission.

STS-107

On January 16, 2003, after two years and many delays, the Orbiter Columbia with seven crewmembers lifted off from Kennedy Space Launch Complex.

Approximately 81 seconds later, at an altitude of 66,000 feet and traveling at 2.5 Mach, a triangular piece of foam about the size of a pillow and weighing 1.67 pounds (one of the two external Tank Bipod Ramps pictured below) broke off. It slammed into the leading edge of the orbiter's left wing at several hundred miles per hour.

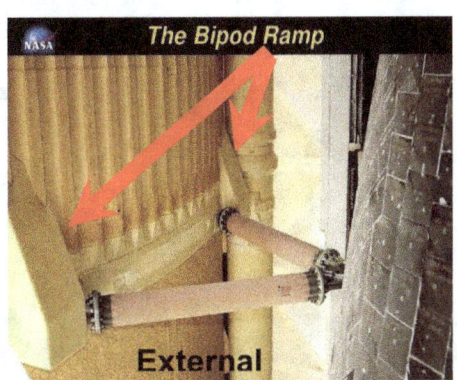

Bipod ramps, one of which hit the leading edge of the STS 107 Space Shuttle

The wing's leading edge was made of Reinforced Carbon-Carbon (RCC) and looked as sturdy as a brick wall. Unfortunately, it was relatively thin, brittle and the hard foam punched a good-sized hole in it. (Please see the image below.) The hole in the wing and the hole in NASA's safety diligence immediately doomed Columbia to disaster before it ever reached orbit.

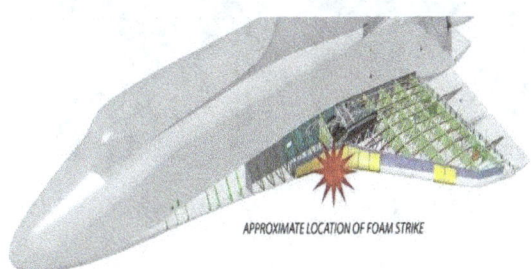

Simulated strike on the wing

The crew could not sense the impact because of the noise and vibration during the first stage of the launch, and they could not see the hole once in orbit because it was not visible from the crew cabin. Had their mission been one that docked with the ISS, it would have been obvious.

Also, they did not have a camera mounted on the end of a robotic arm to inspect their spacecraft, as on many other missions. The ground had images of the strike, but they were not clear enough to define its exact location or the extent of the damage.

However, some engineers realized that the damage could be catastrophic and spoke up. They reasoned that the crew could have performed a spacewalk to inspect for damage. But they were overruled by those who said that foam had come off on previous missions and caused no significant problems. (Equivalent to "We've run this red light many times before and had no problem!") With this feeble approach to crew safety again, the mission ran its course until re-entry.

Crew Before Re-Entry

The bottom row in red shirts is Kalpana Chawla, mission specialist; Rick D. Husband, mission commander; Laurel Clark, mission specialist; and Ilan Ramon, payload specialist. In the top row wearing blue shirts are David Brown, mission specialist; William McCool, pilot; and Michael Anderson, payload commander.

During their mission, the crew enjoyed their time in space and the opportunity to contribute through their work. They also took a picture of themselves (above) that showed they had no pre-knowledge of what awaited them in their re-entry. The role of unprocessed film that contained this picture survived the re-entry and was found in the wreckage debris.

Entry

On the morning of February 1, 2003, Columbia began its re-entry and headed for home on the east coast of Florida. At Entry Interface, an altitude of 400,000 feet (75.76 miles) over California west of Sacramento and at a speed of Mach 23, the catastrophe began to unfold. The ram heating of the atmosphere heated the leading edge of each wing to about 2,800 °F. A small portion of this heated air rammed into the open hole in the leading edge of the left wing and heated the aluminum structure inside like a blow torch. The wing then ripped apart.

Credit: Dallas Morning News
Re-entry of Debris

Like in the Challenger accident, it all happened quickly. The shuttle turned sideways, spun, and broke up as the left wing was destroyed. The re-entry of the Shuttle debris was evident to the naked eye by thousands of observers on the ground as silver streaks and clumps of glowing metal.

The Space Shuttle was an outstanding technical achievement but, as with all complex hardware in extreme environments, its idiosyncrasies must be continuously monitored and analyzed. New engineers and managers and the older, more complacent ones must learn to always pay attention to the sometimes-subtle needs and hazards displayed by the hardware and aggressively take timely and adequate action as required. Even highly subtle hazards do not fix themselves!

As in the Challenger accident, the tragic loss of life in the Columbia accident was again worsened by the required halt in the forward momentum of entering space. Each of the 14 astronauts lost in the last two accidents strongly believed and were flying to add their forward push to this momentum. But in the aftermath of the Columbia disaster, Shuttle flight operations were again suspended for a lengthy period of 2.5 years.

> **Assuming positive results from uncontrolled operations will continue is not a replacement for constant vigilance and appropriate actions.**

Concluding Missions

Finally, on July 26, 2005, Discovery was launched on the Space Shuttle Program's 114th mission. Onboard was a pressurized logistics module, other ISS supplies, and equipment to perform repair work on the ISS. One objective was the flight safety evaluation, and testing of the Space Shuttle, which, unfortunately, began even before the solid rocket motors (SRBs) had stopped burning when a small fragment of tile came loose and was ejected. Then, at about 5 seconds after SRB shutdown, a large piece of debris came loose from the External Tank (ET), which weighed about a pound and was approximately half the size of the chunk of foam responsible for the loss of Space Shuttle Columbia. Fortunately, it did not impact Discovery. About 20 seconds later, a smaller piece of foam broke loose from the ET, did impact the wing of Discovery but did not have enough energy to do any significant damage.

The following day NASA announced that all future flights were to be postponed until the reason behind the foam loss during the powered flight was determined. As the search began, category 5 Hurricane Katrina hit the Gulf coast at the end of August, impeded the work at all Gulf Coast NASA Centers, and established itself as a deadly and the most expensive hurricane in our Nation's history. In December, it was determined that the foam on the External Tank (ET) was loosened during the thermal contractions and expansions during filling of the ET with its cryogenic propellants.

After several further technical and weather delays, STS-121 was finally launched on July 4, 2006, almost a year after the preceding shuttle mission.

The great majority of the 20 remaining missions were devoted to completing the Space Station assembly, including enhancing its science capabilities by installing the US Harmony module in October 2007.

Credit ESA/NASA
Rex Walheim working outside the European Columbus Laboratory

Then came the European Columbus Laboratory in February 2008, the Japanese modules Kibo and RMS in June 2008, and their Experiment Logistics Module-Exposed Section in July 2009. The last shuttle mission to the ISS occurred on July 11, 2011.

During the 30 years of the Space Shuttle Program, which provided 8,280 days in orbit at a total cost of $209 billion, 355 astronauts and cosmonauts (306 men and 49 women) traveled aboard the shuttle on its 135 flights. Some were able to fly many times, e.g., Franklin Chang-Diaz and Jerry Ross each flew seven times on the Shuttle, a world record. The Shuttle's longest mission was 17 days and 18 hours on the STS-80 mission of Columbia, launched in 1996.

All Shuttle missions combined launched 1,597 metric tons of cargo. But how can you picture that amount of mass? (Think of it as 721 Tesla Model S cars.) The Shuttle also returned 6.5% of this total launched mass. (47 Tesla Model S cars). The Shuttle deployed 180 payloads, including many satellites. However, it was expensive to develop and operate ($450 million/mission; at least $7,000/pound of load). Clearly, hardware with lower cost and higher performance is required for our next space workhorse.

The Shuttle and our ability to put Americans in space were permanently grounded with no replacement by the Obama Administration on July 21, 2011, 50 years after we began! The fleet leader, Discovery, had made only 39 flights despite each orbiter's certification for 100 flights. Clearly, for America to continue our initial entry into space, leaders with Vision, Courage, and Commitment are essential. A leader lacking these qualities can bring it all to a halt and significantly add to the total cost when new leaders must make extra investments to get America back on trackIn this case, it was almost an entire decade after President Obama shut down our human space flight before American crews were again launched from American soil. But it was done by American free enterprise—SpaceX—not our government.

America launching Americans comes to Earth for almost a decade!

Final landing of Atlantis and Space Shuttle - July 21, 2011

I won 133 games, lost only two, but now I'm grounded… forever.

NOT FAIR!"

II – Our First Major International Cooperation

Chapter 15
International Space Station

There is no better symbol of international cooperation and what our United Nations should be all about than our International Space Station visible to almost every citizen on Earth. Our partners and we now focus on the growth and utilization of our combined capabilities in Earth orbit.

International Teamwork

Many nations have contributed to the International Space Station (ISS) hardware, staffed laboratory facilities, flight crew, or logistical support. Although the primary research laboratories were supplied by the United States, Russia, Europe, and Japan, half of the 202 countries recognized by the United Nations have participated in research and education on the ISS.

Along with astronauts and cosmonauts from non-European nations, at least thirteen European astronauts have worked in the Columbus Lab who have come from Belgium, Denmark, France, Germany, Italy, Netherlands, Sweden, and the United Kingdom.

The first continuous manning of the ISS began with the Expedition One Crew, which was launched on October 31, 2000, over a quarter-century after the end of Skylab. For over 20 years, humans have continuously lived and worked on board the ISS, and its use continues today. Many international partners have joined in, but it was expensive, and we have covered over 71% of its cost.

In 2005, Congress designated the ISS as a National Laboratory, which expanded its use to non-NASA academic, government, and commercial investigators who could show that they could utilize microgravity for benefits down here on Earth. The legal owners of the station (US, Russia, European Union, Canada, and Japan) are responsible for their respective elements.

Inside the ISS, one excellent laboratory after another is packed with incredible experiments getting better year after year.

Science Laboratory in the ISS

Space, Ever Farther, Ever Faster – Now!

A wide range of different-sized research facilities and companies have leveraged the ISS microgravity to solve complex problems on Earth. Researchers from over 100 countries have used it, and in fact, over 230 astronaut investigators from 18 countries have visited and utilized the Laboratory. Investigations have been performed in many disciplines, including microbiology, human biology, space science, fundamental physics, meteorology, oceanography, geology, astronomy, astrophysics, and solar physics.

The ISS is mankind's first united, long-term stride off our planet.

Budgets

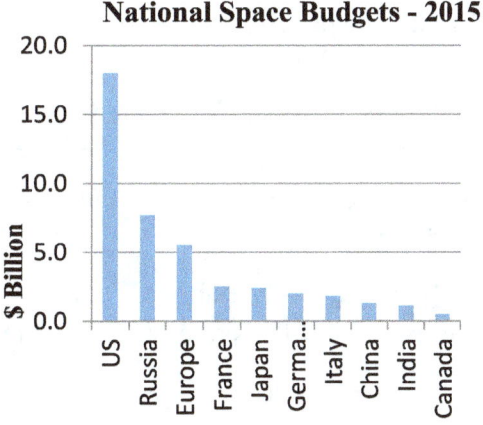

The Center for the Advancement of Science in Space (CASIS), a top-level manager of experiments on the ISS, awarded 45 new contracts and delivered 76 experiments to the station in fiscal 2017. Although the ISS's advancements in science and technology development are most often touted, the "I" in ISS is of equal importance, including international teamwork, contributions, crews, and the unified pride in the entire team's accomplishments. The total space budgets of many of the ISS participants are shown to the right.

America has the responsibility to lead. Certainly, ISS is the largest international research project ever undertaken and, quite likely, the most productive.

China has not been a direct participant in almost every space partnership of which America has been a participant. Their desire to extract far more information and technology from partnerships than what they have ever contributed has limited our interactions with them. However, they have significantly increased their spending on their space programs in recent years, which we have carefully monitored for its magnitude and intent.

Science

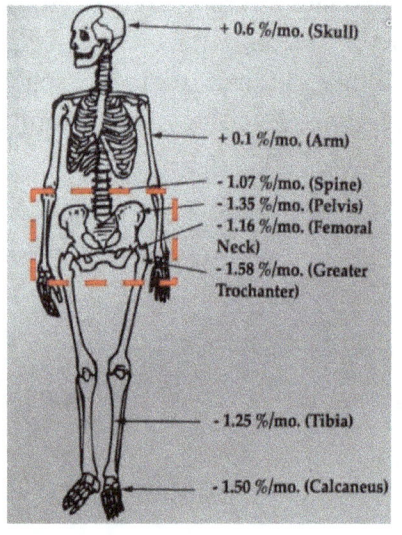

Credit: Caspar Benson/Getty Images

We are continuously increasing our knowledge in areas of medical science, including muscle atrophy, bone loss, biochemical alterations, the bodily fluid shift towards our heads in microgravity, vision changes, and other crew health issues, all of which will enable us to travel farther and longer on future space missions.

Related work includes studies of 3-D human-like tissues and protein crystals. Work is also being done on the physics of fluids and other materials in microgravity, meteorology, astronomy, solar and cosmic radiation, and the quest for antimatter and dark matter.

Bone Loss – The above figure indicates the relative amounts of bone loss from various parts of the skeleton from living without gravity.

Microgravity Countermeasures – The human body tends to lose bone and muscle mass in space, and our success in future long-term space missions depends on their retention. Countermeasures including nutrition (vitamin D), drug use, and adequate exercise are being developed. The benefits of this research on the metabolic processes that lead to osteoporosis apply to humans in space and on Earth.

Drug Micro-Encapsulation – The microgravity on the ISS has been used to help develop drugs encapsulated in small shells (micro-encapsulation) that can then attack many human tumors with chemotherapy.

Microorganism Survival – It has been shown that many microorganisms, or their DNA, can survive quite well in the vacuum and temperature extremes of space. This fact has major implications on possible human contamination of bodies other than Earth, back contamination of Earth from these bodies, and the origin of life on Earth.

Bacteria Virulence – Because of the increased virulence of bacteria in space, more research has been concentrated here that could lead to vaccines to combat these bacteria, such as salmonella.

Robotic Tools – The Canadian-built robot arms for controlling objects in the Shuttle payload bay or external to it have significantly advanced some areas of robotics. Some of these advancements are being translated to much smaller robotic tools of the surgeon, including those used in brain surgery.

Cold Atom Laboratory - This fundamental science experiment was launched to the ISS on May 21, 2018. It created a small cold spot in the station, 0.1 billionths of a degree above absolute zero using lasers and magnetic fields, which were then trapped in place by weak magnetic fields. At an extremely low temperature, atoms become almost motionless and behave as a single wave rather than as individual particles, which cannot be accomplished at the temperatures achievable in the gravity of Earth. Studies of this state of matter yield important information in quantum mechanics' fundamental field. If we want to have the technology that enables tomorrow's inventions and technical strength, we must better understand today's quantum physics.

Flames in the Absence of Gravity - Flames in one gravity and zero gravity, the former is what we are familiar with on Earth that is driven by convection, and latter has no convection, Both are shown below.

Dark Matter - In the field of astrophysics, we seek to understand the characteristics of dark matter, which is about a quarter of all the mass energy of our universe and over five times the mass of ordinary matter, which we used to believe was the only constituent of our universe. Detectors above our interfering atmosphere can only study dark matter's role as a source, absorber, or scattering site of cosmic radiation or antimatter (positrons).

The Alpha Magnetic Spectrometer on the ISS is such a detector. It measures the direction and energy of cosmic radiation and is the most sensitive particle detector ever put into space. Its principal investigator, Dr. Samuel Ting, has previously received a Noble prize in related physics research.

Dr. Samuel Ting

Unexpected Opportunity: Scott Kelly and Mikhail Kornienko

With an eye to the challenges of the future of long-term human spaceflight and spurring applications down here on Earth, a fascinating study of the effects of living in zero gravity and space radiation for 340 days was launched on March 27, 2015. Also on board with Scott Kelly for the total duration of his mission was Russian Cosmonaut Mikhail Kornienko.

In addition to its long duration, this study was exceptional because, while astronaut Scott Kelly flew on the ISS, his identical twin, retired astronaut Mark Kelly, remained on the ground as the control subject.

Mark and Scott Kelly

Both brothers deserve credit for making significant contributions to future human spaceflight. They supported and sometimes endured continuous medical monitoring that was often interesting but occasionally irritating.

<u>General Changes</u> - At mission conclusion, the differences between the two brothers included what we already generally understood: Scott's body fluids immediately shifted upward, and 3 to 4 pounds of fluid were initially lost, muscles weakened, he grew about two inches taller, but he quickly lost his new height upon his return to gravity, his bones weakened from loss of Calcium, and upon return to one gravity his balance was unstable.

Also, eyesight is degraded for some astronauts, some of which may be permanent. Scott's cognitive ability remained unchanged. However, about 7% of Scott's telomeres have longer-term changes of many months that may impact bone formation, immune system processes, DNA repair, Oxygen tissue levels, and bloodstream carbon dioxide levels.

The study also revealed changes in Scott's genes. His telomeres were altered but NOT his chromosomes. His chromosome DNA was unaltered, but its expression was. Telomeres formed at the ends of each 46 chromosomes, 92 in total, protect each chromosome from damage or joining end to end with one another. They partially define how each of our genes interacts with their environment how genes express themselves.

Thus, telomeres are highly important to humans and other species, which depend on chromosomes to sustain and replicate life. Each telomere is made up of a chain of several repeated DNA sequences that shorten with age. What was noted in Scott's telomeres is that they lengthened significantly, although most of them lost their new length within two days after landing. Further study, which involves about 200 researchers, is now in process and will tell what effects persist over time.

When Scott Kelly got the news, he tweeted, "What? My DNA changed by 7%. Who knew? This could be good news. I no longer must call Mark my identical twin brother."

The results of this and future studies will help define the impacts, limitations, requirements, and opportunities for humans on future lengthy space missions like those to Mars. Unfortunately, this study could not include the effect of the type of radiation that exists beyond the protective shield of Earth's magnetic field.

However, about 7% of Scott's telomeres have longer-term changes of many months that may impact bone formation, immune system processes, DNA repair, Oxygen tissue levels, and bloodstream carbon dioxide levels.

We still have much yet to understand.

Total Flight Duration - Although Scott Kelly was on the ISS for 340 days in the mission cited above, he has spent a total of 520 days in space. Peggy Whitson holds the American record of 665 total days in space (1.82 years). Even as we take considerable pride in these accomplishments, we must also recognize that the record for the longest continuous time in space is 437 days and belongs to Valeri Polyakov from Russia. The record for the total time in space is 803 days (2.20 years!), also owned by a Russian, Sergei Karikalan. However, this record is now also in the process of being broken by Sergei's countryman, Gennady Padalka. Intended or not, these records remain a challenge and focus for America.

Earth Views – With our intense interest in images of space objects, we sometimes overlook images of Earth from space. The recent picture of the eruption of the Raiko Ke Volcano on the Kamchatka Peninsula of the Kuril Islands is an example (below).

The two previous eruptions of his volcano occurred back in 1778 and 1924. The eruption shown below occurred on 6/22/19 when an explosion of gas and ash from the 2300-foot-diameter volcano shot up through the overlaying cloud layers. The dynamic nature of this explosion is evident not only in the central upwelling column but also in its capture by the circulating high winds of the surrounding storm, which then carried it east over the North Pacific Ocean. Once the gas pressure in the upwelling had equalized with that of the surrounding gas, the upwelling was quelled, spreading the gases.

Space, Ever Farther, Ever Faster – Now!

Raikoke Volcano Eruption

Completed ISS

The ISS has become an extremely productive facility and improves with each passing year. It provides us the hard-earned base from which to explore, develop, utilize in many areas and expand efficiently.

Credit: ESA/NASA
Handover at the end of ISS Expedition 51
Top - Thomas Pesquet, Peggy Whitson, and Oleg Novitskiy
Bottom - Jack Fischer and Fyodor Yurchikhin

Space, Ever Farther, Ever Faster – Now!

Americans in Space

Who cannot help but smile and take pride in the patriotism of Christina Koch and Nick Hague on the ISS, the only two Americans onboard the ISS on *Independence Day: Christina Koch and Nick Hague on the ISS - July 4, 2019*

Christina continued and on 12/30/19 had flown 289 consecutive days, which broke the world record set by Peggy Whitson for the longest single spaceflight by a female. She continued to fly into 2020 when she performed three spacewalks with Jessica Meir to repair Space Station equipment, also a record but this time for all-female spacewalks—three times!

When Christina returned to Earth on 2/6/20, she had flown 328 consecutive days, which is just 12 days short of the single American flight record set by Astronaut Scott Kelly in 2016. Records or not, Christina has an excellent attitude. "It's not how many days you fly, but what you do with each day!"

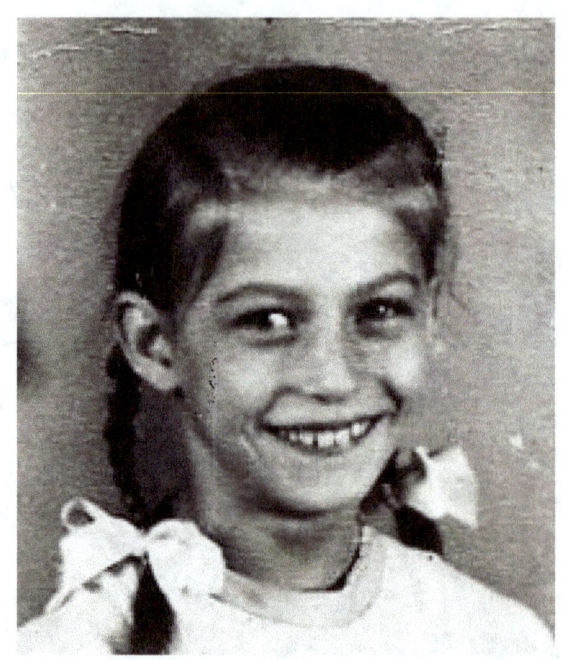

Just as proud as we American girls can be!"

This Cute Little American Girl is now also my wife Julie of 65 years!

✳✳✳ (Green)

Chapter 16
America Struggles in Space

Even with the success of Apollo's Moon landings and Skylab, the advanced capabilities of the Space Shuttle, and the continuous growth of the International Space Station fresh in our minds, America in Space had to struggle to stay alive — let alone to grow.

President Nixon's "Space Doctrine" assured that an American astronaut would never again go beyond Earth orbit for at least the remainder of the 20th century and a good deal of the next. Also, any new, ambitious space plans in Earth orbit or farther out would face major resistance for at least the same period.

Our Reduced Pace

There is no question that America has had some incredibly significant space programs. In first period of 14 years, we had a rush of new developments. In the next period, which is three times longer than the first, we have had fewer developments yet still significant ones.

FIRST 14 YEARS

5 Manned Space Programs

6 Lunar Landings

1st U.S. Space Station (Skylab)

1st U.S. International Space Program (Apollo Soyuz)

1st Large Human International Space Program (ISS)

Skylab 1973-1974

Apollo Soyuz 1975

NEXT 42 YEARS

1 Space Shuttle

1 International Space Station

Commercial SpaceX Rockets

Space Shuttle 1977-2012

International Space Station 1998-present

In 2004 President George W. Bush stepped up with his "Vision for Space Exploration." It planned to retire the Space Shuttle, complete the International Space Station and develop a new launch system able to return to the Moon by 2020 — then go on to Mars. This initiative had: Vision, Courage but, unfortunately, not enough Commitment!

President Bush had the vision and courage to pursue his new initiative, but his team lacked the commitment to see it through. He allowed his Administration to significantly underfund it in its early years, which slowed its development and significantly slipped the date of its first flight. With its fixed annual costs in its government-contractor structure, it became both "over schedule AND over cost."

When President Obama took office in 2011, he hired a blue-ribbon committee that told him, not surprisingly, that the Vision for Space Exploration was "over schedule AND over cost." So, he canceled the program then told NASA to go find an asteroid. At almost the same time, and as previously planned, the Shuttle Program was also terminated.

Regrettably, even though circumstances were vastly different than when the termination was initially planned, the Shuttle was not extended. The Bush administration had made it too easy for President Obama to delete what he did not want to do. The launching of Americans from American soil came to a complete halt, which lasted for almost a decade to 2020!

The days of Vision, Courage, and Commitment vanished and were replaced with… *NOTHING!*

"Devastating and condemns the US to a long downhill slide into mediocrity" - Neil Armstrong

Neil expressed our thoughts well. This one-two punch of Constellation and Shuttle cancellations completely knocked out and ended America's 50-year Golden Era in Space, and the debilitating slide continued. Only the underfunded ISS remained in service. And for almost an entire decade, America was dependent on Russia to put each American into space at the cost of $81 million each, which is four times what it cost a decade previously. Why did Russia charge this high price? *Because they could!*

Our weakened participation also changed the ISS staffing makeup. Although we funded over 71% of the station's cost, we accepted a 50-50 ratio of American to Russian crews; sometimes, it was down to 40-60. Also, outside of the United States, one-third of all the world's launches were Russian, and 40% used Russian engines. Worse, some of our high-priority national programs had become dependent on those Russian engines.

No one planned it, but then, no one stopped it either.

> **But now, despite our many accomplishments in Mercury, Gemini, Apollo, and Skylab,**
>
> ***will America ever commit to human missions beyond Earth orbit again?***

Ed Gibson

Our Decreased Launch Capability

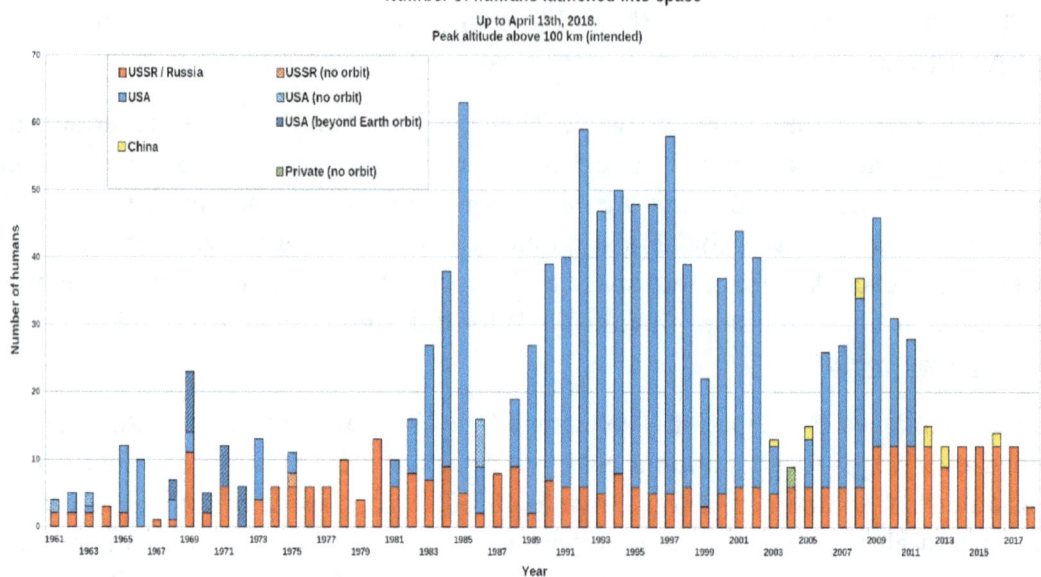

Credit: <u>Billy Meier UFO Research</u>

Summary: The number of launches by the USSR (RED) approximately doubled in 2009. By 2012, the number of launches by America (BLUE) dropped to ZERO. Fifty years after we began, we could no longer launch even a single human into space! Depressing… simply depressing. How did we ever let this happen? Where do we wander off to now?

In the words of our oft-quoted, great American philosopher, Yogi Berra, "If you don't know where you're going… you might not get there."

Fortunately, American free Emprise stepped up in the form of SpaceX and replaced the hitchhiker in the above figure with a real American space vehicle that first flew on May 30, 2020. We all hope and will do whatever possible to ensure that America never lets itself get placed in such a trailing position again!

Leadership: Other than Presidents Kennedy, Reagan, and, more recently, President Trump, we had no Presidential administration that:

1. Initiated MAJOR new space initiatives with Vision, Courage, and Commitment
2. Recognized that the fruits of Apollo are always available with aggressive implementation.
3. Implemented valid programs to return to the Moon to stay then go on to Mars.
4. Reversed the budget erosion.
5. Returned America to a fair level of ISS manning and utilization.
6. Returned America's ability to launch Americans from American soil… until 5/30/2020!

Finally, on Saturday, May 30, 2020, SpaceX launched two Americans to the ISS from American soil on their Crew Dragon spacecraft. Elon Musk was "overcome with emotion" after SpaceX, the company he founded in 2002, became the first commercial company to launch a rocket carrying humans into orbit, ushering in a new era in space travel.

SpaceX's two-stage Falcon 9 rocket with NASA astronauts Robert Behnken and Douglas Hurley aboard blasted off flawlessly on Saturday from Florida's Kennedy Space Centre for the 19-hour voyage to the International Space Station.

Credit: Reuters

SpaceX Launch 2 Astronauts to the ISS

Elon Musk had the grit and determination to endure several launch failures in his initial rocket testing before he was successful. He was not an extremely wealthy man at the time. Although he is not a native American, he certainly exhibited true American spirit of persistence in his SpaceX rocket development.

As of 10/26/2012, Elon Musk had become the world's richest human with a net worth of $290 billion.

III – Cost Reduction

Human can accomplish much more when costs are much less.

Chapter 17. Lowering Cost

Awareness is growing of the major challenges, inevitable opportunities, that we have in Earth orbit, on our Moon, on Mars, and further out. And if we pursue these challenges in ways "not because they are easy but because they are hard," payoffs in our technology, economy, and education will follow as they did in Apollo. One enabling key to it all is lower cost.

Orion Spacecraft

On the Government Side:

NASA and its supporting contractors are working to provide leadership. After the cancellation of the Constellation Program, some components were resurrected by Congress. The Orion spacecraft built by Lockheed Martin for NASA was the first one restored to life.

Credit: European Space Agency
Orion Spacecraft

It has a crew module (Please see above.) that resembles the Apollo Command Module but is larger and can accommodate a crew of six rather than three on low Earth orbit (LEO) missions and four on missions beyond LEO. It is technically far more advanced than Apollo, and the

European Space Agency is building its Service Module; thus, it is now international effort. The system will be helpful for Earth orbital missions and a return to the Moon. The Orion capsule is based on the design developed in the Constellation Program and where it was called the Crew Exploration Vehicle. That design was initiated in 2005, *19 YEARS AGO!*

Credit: Boeing
CST-100 Starliner

NASA contracts Boeing to provide a similar spacecraft, the Crew Space Transportation (CST)-100 Starliner, which will fly unmanned and crewed test flights at about the same time as Orion. It is designed for missions in Earth orbit that include crewed flights to the ISS.

The SLS

There were three CST-100s in production at Boeing's Kennedy Space Center Facility, each designed to fly up to 10 times. Unlike Orion, it is currently not a candidate for cargo resupply or lunar missions. Also, it is different than Orion in that Boeing has invested a considerable amount of its own funds in Starliner and, in that sense, is more commercial. However, neither Orion nor CST-100 are adequate, by themselves, for missions to Mars. Very few people, even astronauts, would agree to gather up five of their closest friends, climb into an SUV, lock the doors, and sit there for six months. For Mars, we need different

Also resurrected after the cancellation of the Constellation program was the Ares launch system initiated in 2005 and now called the Space Launch System (SLS). Its role is necessary: put much mass into Earth's orbit as possible. Its Block 1 configuration will have 17% more trust at liftoff than our biggest booster to da.te, the Apollo Saturn V. the Block 2 availability is even further out and more uncertain than Block 1.

 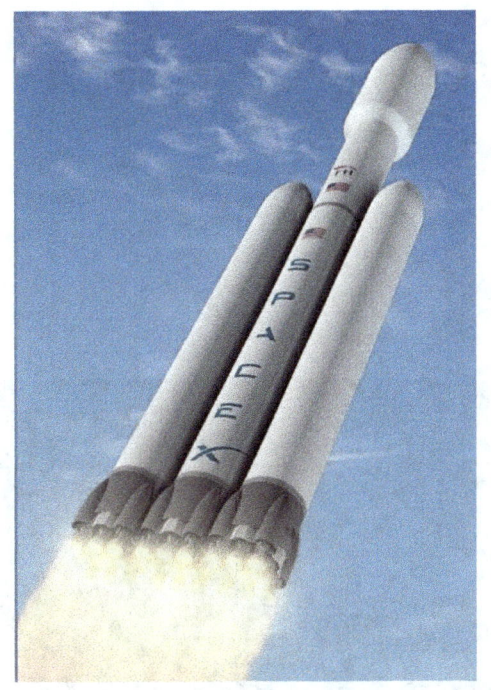

NASA's SLS   Credit: SpaceX

SpaceX's Falcon Heavy

Although NASA's SLS has 40% more thrust than SpaceX's Falcon Heavy, it also comes at over 20 times the cost per launch ($100 million for SpaceX vs. $1.5 billion to $2.5 billion for the SLS). However, SpaceX does not view itself as a competitor to NASA but a partner. Elon Musk, the CEO of SpaceX, believes that the more people or organizations that put effort into a project or program, the greater the chances are of success, which in the end is what it is all about. He also believes that space exploration will become a large public-private partnership.

Elon Musk states that SpaceX's objectives are to "reduce the cost of spaceflight, extend humanity's reach into the solar system and make us a multi-planet species."

In 2008, after three launch failures and risking all that he had on a fourth attempt, his Falcon 1 rocket succeeded. He now has government and commercial launch contracts, and a fantastic record of success. SpaceX itself now has a valuation of $100 billion and climbing.

Elon Musk, the leader of Tesla, Inc. as well as and SpaceX, has become a dynamic trailblazer as he assumes significant American leadership of how "We, the Human, Enters Space".

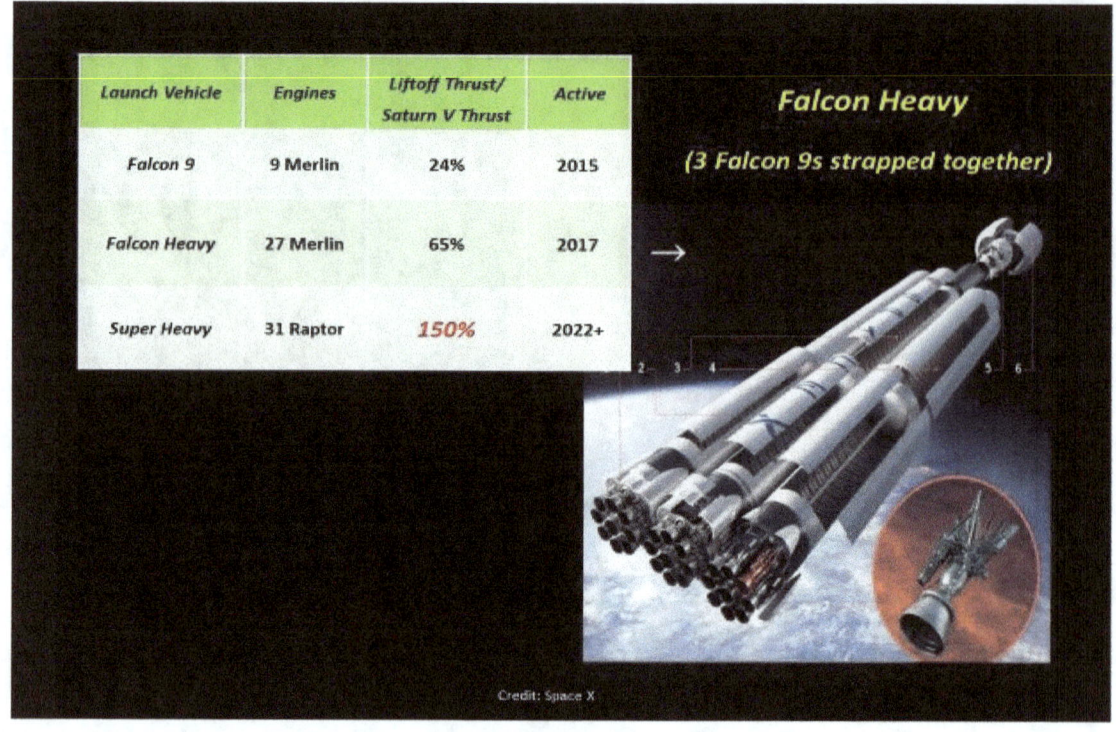

Credit: SpaceX
SpaceX Falcon Family

The base and original member of the current Falcon Family is Falcon 9. (Please see the above figure.) It employs 9 Merlin engines and has a liftoff thrust of about ¼ of the thrust of the Saturn V. Of its first 88 flights, 86 were successful. Attempts to return the booster to the Earth in this same group of fights, 54 have been successful, including those flown on the Falcon Heavy booster. One recently returned booster has flown five times.

Falcon Heavy employs 3 Falcon 9s strapped together to yield a liftoff thrust of about 2/3 of the Saturn V. In its early life, it has successfully flown three times and returned its nine boosters to Earth successfully seven times.

The Super Heavy employs 32 new Raptor engines to produce a liftoff thrust that is 50% higher than that of a Saturn V. The Raptor engine, designed to operate at a chamber pressure of 3,600 psi and even, eventually, slightly above 4,000 psi, has undergone initial testing. SpaceX's longer-term roadmap includes an in-orbit refueling capability to make trips to the Moon possible.

SpaceX states that the requirement of reusability improves the quality of the booster significantly, and those previously flown boosters will prove to be even more reliable than new ones. It must fly successfully at least seven times to be certified for carrying a crew or a classified payload. But its low cost does not mean low quality. Each of its Merlin rocket engines has an extremely high thrust-to-wight ratio, one of the highest ever achieved.

Credit: SpaceX

Falcon 9 booster landing to be used again after launching a payload

SpaceX is in the launch business for the long haul. In 2014 it signed a 20-year lease to use NASA K's launch pad 39A, the same pad from which most of the Apollo missions were launched as well as many Shuttle missions. It is also developing an alternate launch site near Brownsville, Texas, and a Port of Los Angeles facility to construct its 30-foot-diameter second stages, which will be shipped to the Kennedy Launch site via the Panama Canal.

My family and I have been fortunate to receive personal tours of the SpaceX facilities in Los Angeles and Kennedy Space Center. Most impressive was the space hardware returned from space to be used again. Even though it looks ready to go again on the surface, only testing will disclose its actual condition.

The new Block 5 and final version of Falcon 9 to be flown with the crew is intended to be reused ten times without refurbishment. Then, after some refurbishment, it may fly another ten times and repeat this cycle until it has flown a planned 100 times total. So far, many boosters have flown twice, and a few have flown three or four times and one five times.

SpaceX believes that this new Block 5 should meet current commercial and national defense requirements. Also, to reduce costs, SpaceX launches on readiness rather than a fixed schedule whenever possible. An effort to further cut costs is to load the propellant after the crew, enabling a larger amount of propellant to be loaded at a higher density. In mid-2018, NASA approved this proposal.

SpaceX's Falcon Heavy pictured two pages above ("SpaceX Falcon Family") is designed to return all three of its single Falcon boosters to a soft landing to be used again. It has won contracts for military launches and is a prime candidate for the civil market. SpaceX has signed a statement of intent with the Air Force that requires a certification process involving an unspecified number of demonstration flights before the Falcon Heavy could launch expensive classified payloads. However, the financial incentive is high. United Launch Alliance (ULA)'s Delta 4 rocket, which now launches these classified payloads, has only about half the thrust of a Falcon Heavy but costs at least three times more. In June 2018, The Falcon Heavy was selected to launch an Air Force Space Command satellite in competition with the Delta 4 booster by coming in at 37% of ULA's cost. But ultimately, because of the expense of the payloads involved, a launch vehicle's reliability and safety record could be the determining factor for some future missions.

Credit: SpaceX
Falcon Heavy on Pad

On its second flight on April 11, 2019, all three boosters did land successfully, the central one on its ocean platform. On each of its first and third flights, two of its three boosters did make it back to their land platforms.

Credit: SpaceX
Return of Side boosters

Parameter	Saturn V (S-V)	Space Launch System	Super Heavy (with the second stage)
Purpose	Lunar Exploration	Lunar and Mars Exploration	Earth-Lunar, Multi-Planet & Intercontinental Transport, Mars Colonization
Height	365 feet	365 feet	387 feet
Liftoff Weight	6.54 million pounds	5.5 million pounds	9.7 million pounds (1.5 x Saturn V)
Liftoff Thrust	7.89 million pounds	7.2 million pounds	11.80 million pounds (1.5 x Saturn V)
Stages	3	2	2
Payload to Low Earth Orbit	310,000 pounds	209,000 pounds (2/3 x S-V)	330,000 pounds (1.06 x S-V)
Vehicle Only Cost/Launch (2016 $)	$110 million	$2.0 billion +/- 0.5	$7 million with full reusability (6.4 % of S-V)
Project Cost (2018 $)	$42 billion	$23 billion	$ 7 billion (12% of S-V)
Funding	Government	Government	Private and Government

Heavy heat shields will also be useable and rendezvous and docking automated. It is believed by many at SpaceX that the Super Heavy will become economical enough through full reusability and precision landing that it will replace the Falcon 9 and Falcon Heavy vehicles for all launch assignments. No doubt this will be a challenge.

The twenty-nine engines of the SpaceX Super Heavy Booster 4 are seen in the picture to the left as the booster is hoisted onto its orbital launch mount at the SpaceX Starbase in South Texas on 9/8/21. It was topped by a prototype spacecraft. This was the first full stacking of a Starship vehicle.

Credit: SpaceX

When the fully reusable Super Heavy becomes operational in the 2022–2024 timeframe as currently planned, its operating cost will be far less than that of the SLS Block 1B, which itself might not yet even be operational. The Super Heavy will replace the SLS in many of its now planned applications if this is the case.

SpaceX is driving down cost not only by reusing rockets and spacecraft but also by taking propellant into orbit on several separate boosters rather than using just one much larger and much more expensive booster. SpaceX is also planning to produce propellant on the Moon and Mars eventually.

As Elon Musk drives down the cost of spaceflight, his own net worth has rapidly grown even in the era of the Corona virus. In addition to SpaceX, his wealth derives from being an early investor, the CEO and the architect of Tesla, Inc., the electric car company; founder of the Boring Company, a company creating underground transportation capabilities; co-founder of Neuralink; and co-founder and initial co-chairman of OpenAI.

Musk took his spacecraft's name "Dragon" from the song "Puff, the Magic Dragon" because in 2002, when SpaceX was founded, many critics believed that his goals were impossible to achieve. "Falcon" came from the name of the starship in Star Wars, the Millennium Falcon. He is undoubtedly unconventional in his management style and public relations, but clearly, he is also driven, demanding, combative, and focused on reducing space travel costs. He said he "might like to live on Mars, maybe even die there — just not on impact."

Through our free enterprise system and SpaceX's and other companies' enterprising spirit, drive, and achievements, We, the Human, Enters Space with ever longer and swifter strides outward.

The Variable Specific Impulse Magnetoplasma Rocket (VASIMR)

In the pipeline of new engines, there is one that might become one of our major rocket engines of the future because it could enable quicker human flights to Mars and beyond as well as economical cargo transport and space station orbital maneuvering. It is the VASIMR system, an electric propulsion system that has been in development since 1977. Since 2005, Ad Astra has invested more than $40 million in private funds to develop this engine. No doubt it has several significant challenges to achieve the desired performance, but if it does, that performance would be well worth it.

This engine fills the gap between what has been discussed primarily in this book: high-thrust, low-specific impulse chemical rocket engines, which would be required to put VASIMR into Earth orbit, and low-thrust, high-specific impulse ion engines. VASIMR ultimately seeks to provide an engine with both high-thrust AND high-specific impulse performance in the same engine. It is a tall order.

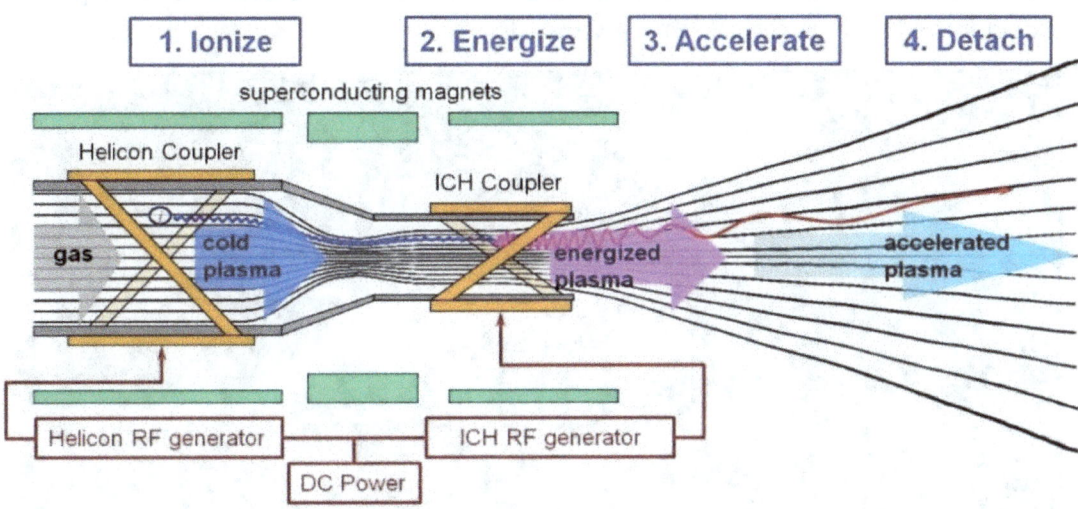

Credit: Ad-Astra Rocket Company

VASIMR Rocket Engine Functional Layout

VASIMR is described below because it might be an important engine in the future and represents the field of electric engines, which holds much promise.

A superconducting electromagnet generates a strong axial magnetic field in a tubular structure containing three cylindrical coaxial chambers: an "ionizer," a "heater," and a magnetic "nozzle." Gas entering the engine (opening at left) is bombarded in the ionizer with radio frequency (RF) electromagnetic energy waves that strip electrons from the atoms. The energized electrons collide with neighboring atoms, causing a cascade of ionization that turns the gas into a plasma. This type of ionizer chamber is also called a "Helicon," referring to the helical RF plasma oscillation modes that resonate and are responsible for delivering the energy from an antenna (or coupler) to the plasma. At 40,000 to 50,000 °K, the initial plasma in the Helicon ionizer is already hotter than the sun's surface (4000°K–5000°K) (5). Fortunately, the magnetic field insulates the hot plasma from the vessel walls.

Following the guiding magnetic field, this "relatively cold" plasma then enters the second stage of the engine, the ion cyclotron "heater" (ICH). In this second section, electromagnetic waves are tuned to resonate with the orbital RF frequencies of the ions as they spiral along the magnetic lines of force; that is, the plasma is very significantly energized. Its temperature can reach millions of degrees, comparable to the sun's interior.

Lastly, the hot plasma now enters the magnetic "nozzle" where the particle spiraling motion is turned into axial motion, a process induced by the gradual divergence of the magnetic field and the requirement of energy conservation. This "magnetic expansion" results in simultaneous cooling and acceleration of the plasma that detaches from the field to provide thrust.

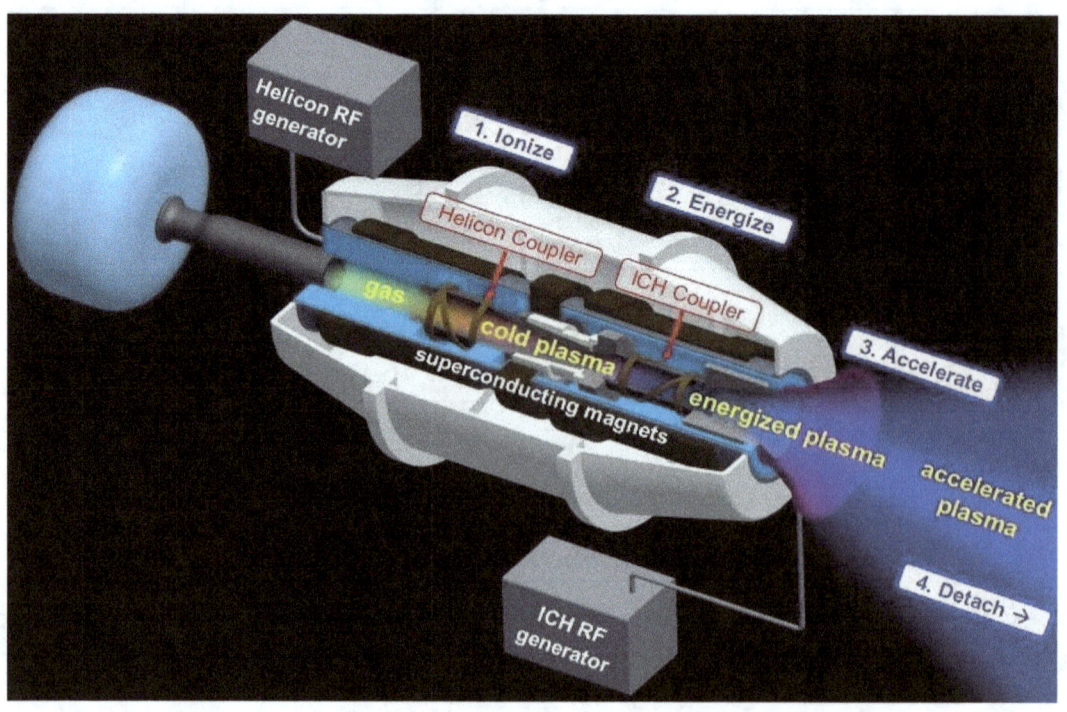

Credit: Ad-Astra
VASIMR Rocket Engine Simplified Physical Layout

The exit velocity of the plasma could be extremely high (50,000 meters/sec). This engine does not use electrodes, so that electrode erosion does not occur as it does in other electrical propulsion devices. The magnetic fields also keep the hot plasma away from the engine's inside walls, whose melting point is well below the plasma temperature. A wide range of gases can be employed as the working gases, including argon, deuterium, Helium, nitrogen, xenon, or some combination of two or more of these gases. (Please see the figure below for a general physical layout of this engine.)

This description is how the engine is designed and desired to work for maximum performance. If this performance is achieved, flight times of two months or less to Mars might be possible for human missions. The heavier support equipment can be launched first to lessen requirements for the crewed ship, followed by a smaller "fast-boat" system that carries the crew.

However, the biggest practical issue in the way is the extreme amount of energy that must be added to the plasma in the second section. A high-power nuclear-electric generator is required, but none now exists that is strong enough and light enough to be considered for flight hardware. For example, for a human mission to reach Mars in 40 days, a reactor and power conversion system weighing less than 1 kg/kW would be required. Present-day, state-of-the-art technology is at 10 kg/kW, so much progress is needed, and development must start in earnest!

In the interim, while nuclear-electric technology matures, Ad Astra contemplates using the VASIMR engine in a host of solar-powered robotic commercial applications in cislunar space. With near-term technology, solar arrays operating at power levels of about 300 kW or less are achievable. They could provide a far more economical and sustainable transportation alternative to conventional chemical propulsion.

Nonetheless, two lesser operational aspects of the VASIMR☐engine are worthy of discussion. First, the plasma exhaust has a rotational component that will provide rotational torque to the vehicle; and second, the magnetic dipole will react with external magnetic forces, such as the Earth magnetic field, and provide an unwanted torque.

These effects could be "zeroed-out" by flying left-handed and right-handed engines side-by-side. This approach is like counter-rotating propellers in aircraft.

Credit: Ad-Astra

Parallel Firing Engines Cancel their Opposite Rotational Velocities

Please see the figure above in which two VASIMR engines are operating in parallel, thrusting a payload.

A second approach to neutralizing the magnetic torque at Ad Astra is the magnetic quadrupole configuration, where an outer "bucking" coil is added to the inner magnet. Recent design studies indicate that, even with the added weight of the bucking coil, the total engine "specific power" in kg/kW is acceptable. At present, the quadrupole architecture is the configuration of choice planned for the company's first mass-produced engine, the 150 kW TC-1Q.

So far, the total system efficiency that has been achieved is about 60%. That is, 40% of the energy lost means that significant energy will result in heat that must be dissipated. The VASIMR engine operates only in a vacuum, has a low thrust-to-weight ratio, and is a candidate for the upper stage of a cargo transfer vehicle, lunar cargo delivery, space system drag compensation, satellite repositioning, and related missions. From the company's inception in 2005, the maturation of the VASIMR technology, then at a Technology Readiness Level (TRL) 2 and a maximum power of 50 kW, proceeded exclusively with private funding.

By 2007, the company had ramped the power to 100 kW in a new test article called the VX-100 operating with a conventional copper magnet. By 2009, a low-temperature, cryogen-free superconducting magnet was designed, built, and integrated into a new test article called the VX-200, which went into operation in Ad

Astra's brand-new vacuum facility in Webster, Texas, about five miles from NASA's Johnson Space Center.

By 2013, the VX-200 had performed more than 10,000 engine firings with Argon at power levels of 200 kW. The performance results from these tests indicated a 60% overall system efficiency (73% thruster efficiency) and a maximum thrust level of 6 Newtons at a specific impulse of 5000 sec. This corresponds to an exhaust velocity of 50,000 m/sec, or about ten times the exhaust velocity of the best chemical rocket.

In 2015, following these results, NASA awarded Ad Astra a nine-million-dollar, three-year contract (extendable to five) to partially support the continuing development of the VASIMR engine in competition with two other electric propulsion technologies. An electrodeless Lorenz force accelerator known as (ELF), studied jointly by Mathematical Sciences Northwest (MSNW), the University of Washington, and the Hall effect thruster, a low power-density Russian variant of the traditional ion engine, which Aerojet Rocketdyne (AR) was trying to evolve beyond its low power range and into the 100-kW scale. All three technologies were given a goal to demonstrate 100 continuous hours of thermally stable operation of their test articles at 100 kW.

Unable to achieve that goal, both MSNW and AR voluntarily exited the competition in contract years 2 and 3, respectively. Ad Astra's entry, the VX-200SS, remained in the competition but, at the end of the third year, required modifications to the engine that would extend the work beyond the contract. To enable the work to proceed, at their own expense, the company requested, and NASA granted, a no-cost extension of the contract to June of 2020.

Testing high-power plasma rockets is complex, not so much by the rocket itself, but by the demands it places on the laboratory infrastructure, including the exceedingly high vacuum system throughput and the complicating and contaminating effects of the chamber walls on the performance measurements. Knowing these growing limitations and the accessibility of the International Space Station as a National Laboratory, the early VASIMR team, even before the founding of Ad Astra, proposed to NASA to test the engine on the ISS. Aware of the limited power capability of the orbital complex, the team also offered for the firings to be done from a battery bank that could be comfortably trickle-charged by the ISS solar arrays. However, while a NASA study concluded that such a test was technically feasible, the integration costs estimated by the Boeing contractor were prohibitively high.

Though NASA is noted for its advanced technologies, it can also resist the positive acceptance of innovative technologies that infringe on the turf previously hard-won by other existing but now stagnate technologies. Regardless of the turmoil within the United States surrounding some recent technology acceptances, China and

India, among other nations, are developing their space capabilities rapidly and not yet so bound by tradition.

Considering the early years of America in Space, there is no restriction on or substitute for aggressive application Vision, Courage, and Commitment.

The main driver behind the VASIMR development has been Franklin Chang-Díaz, a remarkable individual. He grew up in Costa Rica, immigrated to the United States while still in high school, spoke little English but wished to become an astronaut. His intelligence, hard work, and calm nature separated him from thousands of others with similar dreams. With persistence he earned a BS in Mechanical Engineering, then a doctor's degree in plasma physics and nuclear fusion from the Massachusetts Institute of Technology in 1977, the same year that he became an American citizen. In May of 1980, after his second try, he made it into the Astronaut Program. With the same drive and tenacity, he flew on the Space Shuttle seven times, equaling the world record set by Jerry Ross. As he was doing all this, he developed his idea for a new rocket engine, the Variable Specific Impulse Magnetoplasma Rocket (VASIMR) engine. He obtained his first of three patents for it in 1979. He retired from NASA in 2005 then founded and is now the current Chairman and CEO of Ad Astra Rocket Company.

Could VASIMR be part of our future in space? One way or another, with individuals like Franklin Change-Díaz in the lead, there is a good chance that it will.

Broader Commercial Space Launch Market

To assist commercial space concerns, on 5/23/18, the US Commerce Department created the Space Policy Advancing Commercial Enterprise (SPACE) Administration to "support American companies so they could compete on a level playing field." It was designed to be a one-stop location for private space flight regulations.

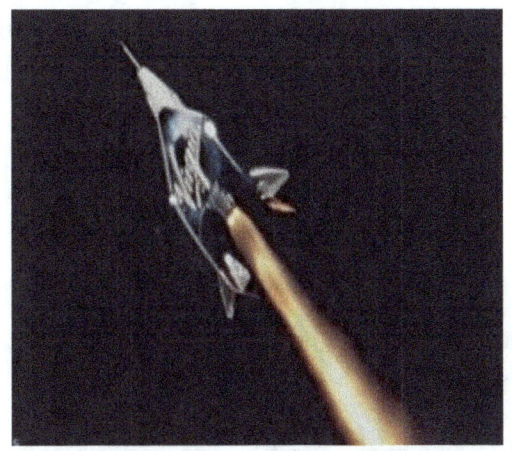

Credit: CNBC.com

Virgin Galactic Spaceship Two Flight on 12/13/18

Credit: Blue Origin

Blue Origin's New Glenn

Other companies and individuals are also investing billions in the space industry. They are not starry-eyed investors, but businessmen focused on making a profit. In 2017, $2.5 billion was invested in start-up space companies, with the number of deals completed increasing by 70% over those in 2016.

Two of the major new companies have been started by billionaires. Initially, Elon Musk invested $100 million of his own money in SpaceX before securing some outside funding. Jeff Bezos, who started Amazon, sold about $1 billion of his stock each year to fund his Blue Origin space company.

Bezos' reusable New Shepard suborbital rocket has flown several times and could fly tourists early the 2020s. His larger orbital rocket, the New Glenn, will have a liftoff thrust of 3.85 million pounds, a first stage that also re-lands and can carry up to six paying passengers or other payloads into Earth orbit. Blue Origin is planning to also certify this New Glenn system for national security missions. But it will not fly until sometime in the 2020s.

Blue Origin's most powerful New Armstrong rocket will follow. New plans could include robotic flights to the lunar surface using its Blue Moon launch and landing systems, although they are not yet in development. Blue Origin still promises to be a strong competitor, which is plainly behind SpaceX in the number of missions flown. Jeff Bezos also believes we will spread outward from Earth at an increasing rate when low-cost transportation is available.

Blue Origin is completing a large factory in Kennedy Space Center's Exploration Park to manufacture its two-stage New Glenn super-heavy launch vehicle. (Please see next page) In addition, a new rocket engine plant is being constructed in

Huntsville, Alabama, which will refurbish and utilize the same Marshall Space Flight Center's engine test facility that once tested Saturn and Space Shuttle engines. Its customers will be Blue Origin and the United Launch Alliance.

Credit: Blue Origin, New Blue Origin KSC Facility
Recently completed Blue Origin's New Glenn Factory
near the Kennedy Space Center

Credit: SpaceX
Falcon 9 (20 missions)

Credit: Northrop
Northrop (12 missions)

The second to last SpaceX mission (CRS-19) launched on 12/5/19 from Space Launch Complex 40 at Cape Canaveral Air Force Station, Florida is seen above on the left. It was the third use (second reuse) of the same SpaceX Dragon Spacecraft in this program. Northrop Grumman (NG) Company launched the last of its CRS-1 12 missions (Antares rocket/ Cygnus spacecraft) to the ISS on 11/2/19 from Wallops Island, VI. This system was the first commercial spacecraft to provide a boost to the ISS (Please see above right.).

In 2019, NASA announced the winners in the CRS-2 Competition: SpaceX (Dragon spacecraft launched on their Falcon 9 rocket), NGIS (Advanced Cygnus spacecraft launched on an Atlas-V booster), and Sierra Nevada Corporation (Dream Chaser spacecraft launched on a United Launch Alliance Atlas 5 rocket).

Government or Commercial?

In the past, "We, the Human, Enters Space" has been primarily led by our government that supplied the funds and schedule and dealt with the wins and losses; Mercury, Gemini, Apollo, Skylab, Apollo Soyuz, Space Shuttle, and the International Space Station have followed this pattern. But can it continue far into the far future?

Clearly not!

Initially, our race with the Soviet Union was the key motivator. Once that incentive faded abruptly after Apollo 11, so did the near-unanimous political support for new aggressive space initiatives until the government's enthusiasm, financial support, and unified leadership have dropped to where it was just a few years ago.

Needs and opportunities in space are rapidly expanding. The government's available budget and normally unhurried bureaucracy are not.

Fortunately, private enterprise is reducing the cost of transportation and a growing number of other hardware items and operations. Profit opportunities are appearing. According to SpaceTec, a German consulting firm, over 60 venture capital firms have also invested approximately $3 billion in space companies between 2010 and 2016. Estimates of our total space economy growing to over $5 trillion by 2025 have been made, which is highly ambitious.

Overnight success is rare, but investment, product development, and sale of ever-lower-cost, reliable services produce significant returns for an increasing number of industry pioneers. As the numbers build and some business is generated within the industry itself, the industry edges towards exponential growth. However, bubbles in this industry, as in most other industries experiencing rapid growth, are also a fact-of-life. For example, because of the ever-decreasing cost to enter the small launch business, many firms with this objective have sprung up, including low-cost firms from India and indirectly from China. Supply is getting ahead of demand, and many hopeful new small launch companies will fail. Fortunately, however, as launch costs continue to fall, both overall demand and supply will continue to rise in the long term.

New or current elements on the ISS can become commercially focused and operated by private enterprises that contract with current and future users, including NASA. Axiom Space and Bigelow Aerospace (below) are two companies interested in attaching a commercial lab to the ISS and eventually becoming independent free-flying stations.

ISS Modules with inflatable Bigelow Module at Upper Left

But where is NASA now in meeting our ISS international responsibilities of ensuring that scientific and development opportunities continue to be available? Certainly, the disastrous decisions made in 2011 to abruptly end the Shuttle and Constellation programs with no suitable replacements available for many years have made this task much more difficult. In the fall of 2019, America's agreement with Russia's Roscosmos to carry astronauts to the ISS expired, which was America's only source of crew transportation. Fortunately, on May 31, 2020, SpaceX launched a crew on its Falcon 9 spacecraft to the ISS. Will Boeing's Starliner or Sierra Nevada's Dream Chaser be ready and man-rated soon?

Then there's the Orion Spacecraft and the SLS booster, which, even if ready by 2023, would be too expensive for most users. And the rate of development of Orion has been anything but swift. Its development started in 2005 in the Constellation program; by 2020, 15 years had passed.

Finally, the American commercial space system, SpaceX, ended our draught in our ability to launch Americans from American soil to the ISS. This depressing drought lasted for nearly a decade.

One proposal for a relatively rapid transition of our space station activities is for a private enterprise to quickly assume responsibility for operation with an annual subsidy of approximately $2 billion from NASA; NASA would then have free access to part of the facility for their own research and development with the private enterprise charging others for the use of the remaining ISS capabilities. Alternatively, an American commercial concern would assume NASA's current operational and maintenance roles and receive fair compensation from the U.S. and international partners who, in turn, would charge for the use of their individual part of the facilities; that is, operate the station as a business. The transition would be complex and unlikely to happen. However, a future station initiated as a business could operate in this mode. The cost will always be a key driver.

Bigelow Aerospace is a commercial concern that might construct and operate such a station as a private initiative (Please see below).

Credit: Bigelow Aerospace

Space station made of Bigelow A330 expandable modules with the same total volume as the ISS

Eventually, the government will return to leading space development efforts with little immediate commercial return, like the early exploration of our Moon, Mars, and other planets and their moons. The commercial sector will invest in more routine support services or products that promise a relatively near-term financial return or in spin-offs of already-developed capabilities that can lead to a significant future return.

This pattern is not new; we've seen it before in many industries, including aviation. First, the government leads exploration and capability development, some of which industry adapts to profit-making opportunities. Eventually, profit can drive a large segment of all evolution.

re`Also, in the not-too-distant future, it is distinctly possible that stations from other nations might take their place in Earth orbit. China (below) is one possible nation that could have a facility in orbit when we plan to terminate ISS operations. Because of security concerns, we have barred the Chinese from using the ISS or sharing any of our technologies. Might they likewise reciprocate if we find we have no other place to carry out our needed research beyond ISS termination? This is a situation to be avoided.

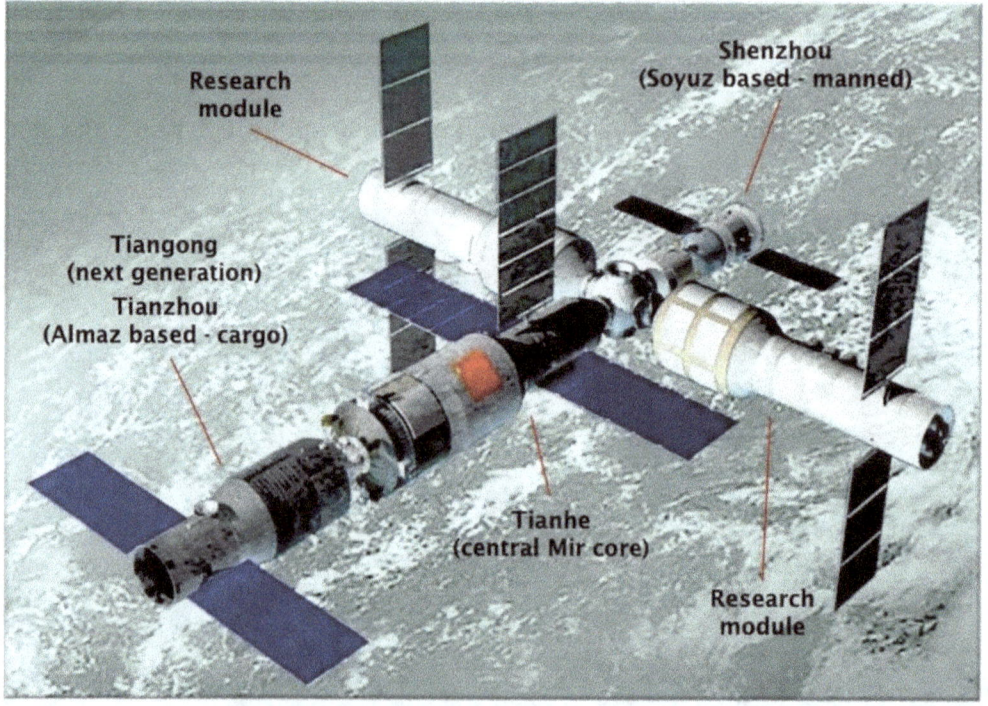

Space, Ever Farther, Ever Faster – Now!

The ISS – an American workhorse in space

Space is becoming more international, diverse, and competitive.

Can American free enterprise continue to give us the edge?

"You bet it can.

Save me a lunar suit!"

IV — Where to First

What should Human's first nearby but permanent destination be after finally escaping Earth?

Chapter 18
Moon or Mars?

For almost a half-century after Apollo 17 returned, we have made no real start of an end-to-end programmatic, commitment (budget, design requirements, total schedule, contractors, etc.) to either return to the Moon or move on to Mars. We tried to initiate Apollo-type programs to both destinations but have always fallen short. With each change of intended direction, we have lost time, opportunities, investments

As many people have correctly advocated, it now appears that our return to the Moon is easiest and should be first up. We have much to learn about not only an efficient lunar return but what lunar resources could enable more efficient explorations of the rest of our solar system. We must regain our ability to take on new engineering challenges of this magnitude. But who will do it?

In general, the two major options are government and American free enterprise. Will a future president come forth and echo the words of JFK 60 years ago?

Space, Ever Farther, Ever Faster – Now!

Credit: George Bush-whitehouse.archives.gov

Let's look at it in picture form. First, we have our strong government space lion as we seek to move outward to the Moon, Mars, and beyond!

NASA, our strong, powerful Space Lion, King of the space jungle, now has more capable people than ever. They are just as resolute and motivated but better technically trained and in possession of far more advanced technical tools than those young engineers who took us to the Moon.

NASA – King of the Space Jungle

But look what we have done to our poor space animal. We put steel chains around its neck!

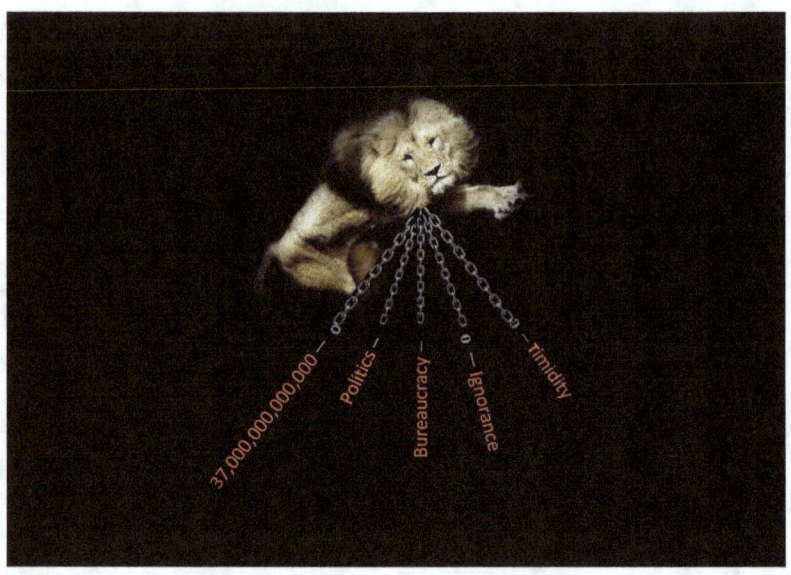

King… in Chains

1. Our 37-trillion-dollar National Debt certainly discourages the initiation of MAJOR new space programs dependent on government discretionary spending.

2. Politicians in Washington who must initiate action but now seem to agree on only one thing — disagree on everything — and often yield to pressures to put money chiefly into those areas that maximize their chances for reelection. Perhaps even worse, America's direction in space most often changes substantially when a new administration with an opposing political party comes into power. Considerable waste, frustration, and misplaced loss of confidence in NASA usually follow.

3. Bureaucratic Processes in the Administration, once the efficient route to a goal, have in many situations remained as the only goal… just like what has happened in many large, aging, and now bureaucratic organizations.

4. Ignorance on the part of most political leaders that the benefits, like we achieved from Apollo in our technology, economy, and education, are always available at any time if we aggressively pursue major space challenges "not because they easy but because they are hard." Undoubtedly, education in America has developed some shortcomings over the past decades, and the stimuli and challenges of space exploration can again be of benefit.

5. Timidity on the part of national leaders to act with real courage and commitment in addition to making speeches focused on heart-warming vision.

Space, Ever Farther, Ever Faster – Now!

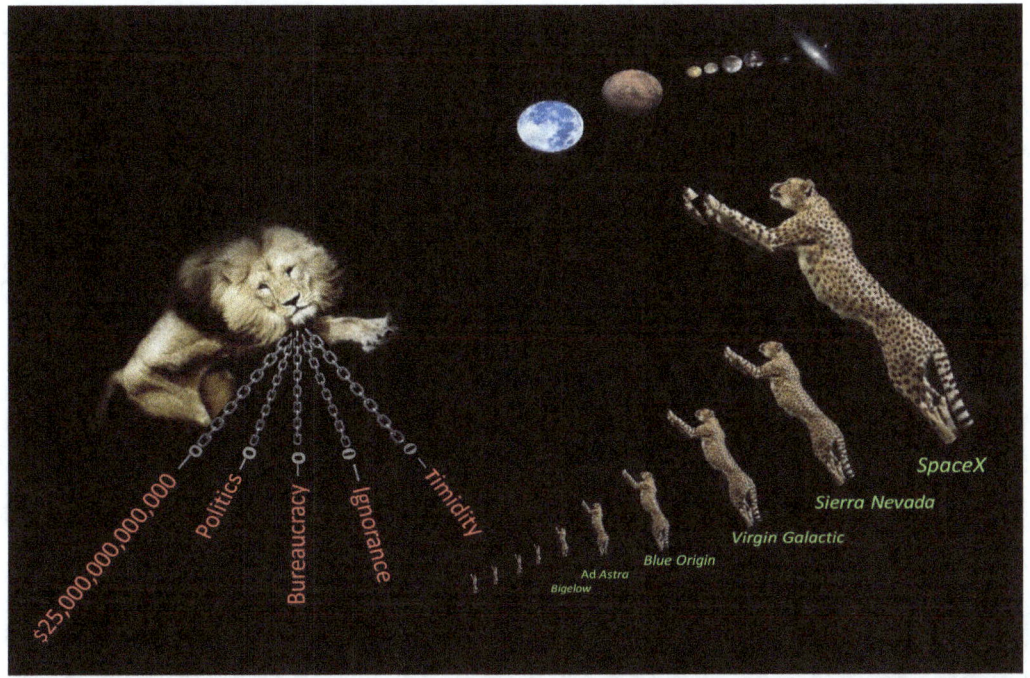

The Space Cheetahs are Moving In!

On the other side, the private sector side, we have many Space Cheetahs, the fastest and most agile animals in the space jungle. They are strong, capable, unrestrained, and highly motivated like SpaceX, Sierra Nevada, Virgin Galactic, Blue Origin, Ad Astra, Bigelow Space, Moon Express, Astrobotic Technology, and many others. They aren't chained down like our Space Lion, and many can also bring some of their own funding and the ability to earn more.

Clearly, the net result will be a combination of both government and industry. But American free enterprise will continue to grow, sometimes hindered, and sometimes assisted by the government, but eventually, it will become the dominant force like what see in the word of aviation today.

American Private Enterprise is Moving In

Of course, this picture leaves out the third large capability that falls between the two extremes, the traditional *large*, competent aerospace companies, whose work in space has been almost totally financed by the government but who brought our technologies and capabilities from where they were in 1961 to where they are today. A few of these companies are Boeing and Lockheed Martin (United Launch Alliance), United Technologies, Northrop Grumman, Airbus, General Electric, Rolls Royce, Honeywell, BAE Systems, L-3, General Dynamics, NGIS, Aerojet Rocketdyne, Bombardier and Harris, and Teledyne.

193

An indication of how the competition between these traditional and the new entrepreneurial companies might work out is found in two contract awards by the Air Force to launch their Global Positioning Satellites (May 2016 and March 2017). Against the highly capable United Launch Alliance, formed by Boeing and Lockheed Martin, SpaceX won by coming in at half their price.

Although hard to believe, eventually, SpaceX will shift into the role of a traditional company as some other young upstarts take over the role of the cheetahs. But now they certainly do not lack enthusiasm.

SpaceX employees celebrating a Crew Dragon as it departed their Hawthorne factory for its Florida launch site where it resumed launching humans to space from America!

Chapter 19
Why?

A question once posed by President Obama is, "Why go back to the Moon? After all, we have been there before."

Let me do my best to provide a few short answers.

1. Often dismissed or ignored but still very real and relevant general answers include our natural human quest to physically explore and understand the world in which we have found ourselves. Fifty years ago, we examined our closest neighbor up close, our Moon. Now we are also looking farther out. Secondly, what initially drove us to enter space, but now deemphasized, is the need to India, and other technically competent nations are ambitious and looking out right alongside us. In the eyes of the world and we Americans at home, we need to re-establish ourselves as a nation of space visionaries, pioneers, and leaders. After our early termination of Apollo before the greatest returns were realized and the termination of our ability to launch crews from America, we need to establish that we can fully meet what commitments we have made. These demonstrate technical superiority and international leadership. China, Russia, challenges are upon us; they are real and can't be dismissed with a diversionary wave of the hand.

2. Although many space workers and followers would like to go to Mars immediately, two generations have passed since we built up our lunar capabilities. Thus, jumping straight into an even more challenging human Mars program without the proper funding, government commitment, and political support that we had at the beginning of our Mercury, Gemini, and Apollo programs would be unwise to be pushed too fast, perhaps even fatal. The new and just initiated Artemis Program is wisely going to the Moon before Mars.

3. A positive tone was taken in 2018 by the new NASA Administrator Jim Bridenstine when he stated, "We are not going back to the Moon, we're going forward to the Moon. We've already left the flags and footprints, and we will stay this time. We are looking at the Moon in the context of our following missions to Mars."

4. Logistically, the Moon is a natural physical steppingstone to Mars and farther out. It provides a base camp for further explorations of our Solar System as well as a natural focus for further development of our required space flight systems and our understanding of the human physiology of

extended spaceflight. As a steppingstone, it can store equipment for to-and-from-flights to Mars as well as a reconditioning site for crews returning from Mars.

5. The Moon provides water resources needed to support ongoing explorations. It is estimated that there are hundreds of billions of tons of water in the form of ice at the lunar poles, especially the south pole. It can be used to drink, breathe once its Oxygen is separated out, and make rocket propellant composed of Oxygen and Hydrogen. This propellant enables missions to and from Earth to resupply consumables and switch out of crews and a flight from place to place across the lunar surface. Harvesting lunar water also helps prepare us for missions farther out: water is about 1 mile below the surface of Mars and available on Enceladus, a Moon of Saturn, and on Europa, a Moon of Jupiter.

6. Scientifically, although Apollo significantly advanced our understanding of lunar geology and our solar system, it was but a start. There is much to be learned from additional focused studies; for example, studies of the central peaks of impact craters like Tycho, Copernicus, or Tsiolkovsky to find materials from asteroids that caused these craters, thereby learning more about our early Solar System.

7. Lastly, the Moon is also a source of another form of energy, Helium-3, a significant potential source of power in terrestrial nuclear fusion power plants and nuclear rocket engines to shorten interplanetary travel times. (2) The Helium-3 atomic nucleus is a light, non-radioactive isotope of helium with two protons and one neutron. Nuclear fusion powers the sun and stars, and its use on Earth for power generation leaves no radioactive remains.

When Helium-3 is fused with deuterium, the reaction will produce energy, normal Helium, and a proton. Nuclear fusion reactors using Helium-3 could efficiently provide nuclear power with virtually no waste, no radiation, and no emission of greenhouse gases. Its presence has been verified by examining Moon rocks returned in Apollo. If the soil from the Moon's surface is heated to around 700 degrees centigrade, its Helium 3 can be extracted as well as water, which results from Hydrogen reacting with oxides both of which are also in the soil.

The value of using Helium-3 in nuclear reactors for power with no nuclear waste has been recognized and promoted by Gerald Kulcinski, the director of the Fusion Technology Institute at the University of Wisconsin-Madison, and other physicists (2) and by Harrison Schmitt, Apollo 17 astronaut (3).

It is estimated that there are over a million tons (two billion pounds) of Helium-3 on the surface of the Moon down to a depth of a few meters. There is more energy stored in the lunar surface Helium-3 than in all the fossil fuels ever identified on

Earth. About 40 tons of Helium-3 could power the United States for a year. Thus, Helium-3 has a potential value of approximately $3 billion per ton. The corresponding value of all the Helium-3 on the Moon calculates to be approximately three quadrillion dollars (120 times our national debt). Obviously, there is much uncertainty in these numbers, but their size does indicate the gigantic magnitude of the opportunity. In addition, in the process of energy generation from Helium-3, Hydrogen, water, and compounds of nitrogen and carbon are also generated as by-products. (3)

It has been projected that there will be at least an eight-fold increase in the world's energy demand by 2050. (2) Contributing factors to this growth are the increase in world population from its current 7.8 billion to approximately 10 billion and the world's increasing dependence on technology, including the currently underdeveloped populations of our world. Putting increasing demands on our fossil fuels, nuclear fission of uranium and other heavy elements, wind and water energy, other current sources, and conservation will be far from adequate.

Utilization of the energy provided by Helium-3 will likely not just be novel but become an absolute necessity. (2)

The rewards from using Helium-3 are great, but so are the many challenges, including establishing economical, reliable transportation to/from the Moon, creating lunar mining and Helium-3 extraction capabilities, and developing economic, commercial fusion reactors.

Once we rationally select our strategy and techniques, are we ready to return to the Moon? When Apollo 11 launched, we were as technically prepared as required. For comparison:

Capabilities in Apollo Versus Those Available Today	
Capabilities Demonstrated on Apollo 11	Capabilities Today
Required booster capabilities (1)	Developed for Artemis:
Required information technology (2)	Far superior to Apollo's capabilities
Required supporting facilities	Some definition and development required
Trained ground personnel	Can be made ready relatively quickly
Trained flight crews	Can be made ready relatively quickly

| Adequate financial support | NO |

Situation Today:

1. Orion and SLS are df over the original schedule and cost.
2. Lunar lander and complications introduced using the Gateway strategy have led to additional schedule and budget issues.
3. The major strides made in the growth of our information technology in the last 50 years have provided far greater capabilities in our design processes and analysis, system testing, nominal inflight system monitoring, and control, automated detection of and responses to technical faults, major system reconfigurations as required and communications.
4. The bright light on the horizon is the infant Artemis Program to be discussed shortly.
5. Clearly, if we are serious about returning to the Moon any time before the end of this decade, rational technical focus and adequate national financial support must remain on track. The private entrepreneurial sector can certainly help with their lower-cost capabilities and focus on near-term results. Still, the government sector must continue to provide strong, rational leadership and financial support.
6. As the financial potential of Helium-3 and other minerals lying in wait on the Moon becomes increasingly obvious, willing participation will arise from many areas, including our American government and industry and international investors and scientists.
7. Budget: NASA's 2024 budget is 24.7.6 billion, numerically the highest budget they have received (061% of our total national budget, 1.46% of our national discretionary budget and an increase of 2.06% over 2021). Fortunately, President Bidden has continued to support NASA as did his predecessor.

Standing next to the President on his immediate left are astronauts Harrison Schmitt and Peggy Whitson (2019 budget signed in December 2018)

Although our government has committed to a Moon's first future of space exploration, a government-dominated approach will be challenging to maintain because of the factors presented in the figure on page 202 ("The Space Cheetahs are moving in!"). As we are about to see, the current government approach to a lunar return that requires an orbiting space station first (a middleman operating as a gateway), which is far less efficient than a moon-direct approach that focuses on immediately landing on the Moon and getting to work where it counts. It is the author's view and that of several others that this moon-direct approach is more efficient and cost effective.

Commercial companies can move rapidly, reduce program costs, be flexible, mine minerals cost-effectively, and expand our economy on and off Earth. These companies include SpaceX, Blue Origin, Sierra Nevada Corp., Lockheed Martin, Boeing, and NGIS. In addition, in 2018, NASA added the abilities of nine more companies specifically to aid in the design construction of systems and make surface scientific explorations even more effective: Astrobotic Technology Inc., Deep Space Systems, Draper, Firefly Associates Inc., Machines LLC., Lockheed Martin Space, Masten Space Systems, Inc., Moon Express and Orbit Beyond. Free enterprise competition should significantly reduce costs. Independent of the government, for example, could competition materialize between Elon Musk, Jeff Bezos, and many other entrepreneurial companies in a race to the lunar surface to its resources?

On January 2, 2019, China soft-landed their Chang'e 4 spacecraft on the far side of the Moon in the Von Karman crater at the south pole, one of the largest craters in the solar system. It carried their Yutu 2 lunar rover, which was immediately shut down to protect it from temporarily high temperatures. A precursor rover, Yutu 1, was placed on the lunar near-side at the end of 2013 but failed to awaken when instructed to do so.

Credit: ASU - NASA's Lunar Reconnaissance Orbiter - On Moon at 45.46° South, 177.59° East

Credit: CNSA
Deployed Yutu 2

However, the Yutu 2 rover ("Jade Rabbit") did awaken when instructed and initiated the operation of a suite of international scientific instruments, including those from Germany, Sweden, and Russia. They have stated that their future lunar science programs will also contain considerable international collaboration. China has also indicated interest and performed early work on obtaining energy from Helium-3h and harvesting lunar minerals.

Once deployed, the Yutu 2 Rover turned around and took a picture of the spacecraft that brought it to Moon's far side, the Chang'e 4.

CNSA/CLEP
Chinese Spacecraft, the Chang'e 4, on the Moon

The question arises, "Who owns the Moon?" Is it to be divided between people or nations who get there first and plant their flags on its soil?

We thought the answer was clearly stated over 50 years ago verbally by Neil Armstrong, "One small step for a man, one giant leap <u>for mankind</u>."

A plaque on one landing leg (right) states, "Here men from the planet Earth first set foot on the Moon. July 1969, A.D. We came in peace <u>for all mankind</u>."

The 1967 Outer Space Treaty was consistent with this intent. In fact, we have physically shared the moon rocks and data returned with all interested nations.

Plague left on landing stage of Apollo 11 clearly states our intentions

The United States believes that anyone can go to the Moon and own what they can bring home, just like the law of Earth's oceans. That is, the Law of the High Seas states: "No one owns the seas, just the fish they land in their nets."

Good. Done!

No, wait... not so fast.

Nations like Russia, maybe Brazil, Belgium, and others, believe that "The Moon and its resources belong to all nations of Earth, and all values extracted must be divided between all of these nations." Before it is practical to tap lunar resources in a big way, this issue must be resolved!

Another way exists of obtaining water than the direct melting of water ice. Harrison Schmitt (3) stated, "With nuclear energy of boot-strapped fuel-cell power systems, water can also be produced anywhere on the Moon by heating the regolith, so Hydrogen reacts with Oxygen bearing oxides and silicates to produce water. Early trade studies are needed to determine whether one or the other or both approaches would be desirable." (4)

Water ice can be used for drinking and broken down into Hydrogen and Oxygen to provide a breathable atmosphere and fuel for rovers and rocket vehicles, which would enable excursions to other parts of the lunar surface. Small nuclear reactors could also enable energy.

How?

When both humans and robots are available on the Moon or in lunar orbit, we must choose which one to employ for any operation to be performed. Do we let the human do it thoroughly, put the human in control of only some of the operations directly or through telerobotics or just turn the robot lose to operate freely within a defied scope?

Need for human judgment, cost, accessibility, time constraints, job importance, safety, and even politics are some of the many factors that inherently go into a choice. Although today's robots have an ever-expanding capability, the returns from the Apollo missions clearly demonstrated the high value of humans on the spot to apply their judgment, operational flexibility, and physical abilities to surface explorations.

A more controversial and highly significant choice is also now at hand; logistically, how will we perform lunar exploration and utilization to maximize our return from our opportunities and limited funds available?

The robots that we send or take to Mars would involve the same variables but emphasized differently.

NASA's current planning is largely focused on a lunar orbiting Gateway Space Station, another space station to be constructed in lunar orbit.

Lunar Gateway Space Station

A two-stage spacecraft will leave the Gateway to land on the Moon and descend to a low lunar orbit where a descent module will separate and land. When a lunar traverse of up to two weeks by up to four crew members concludes, the descent module will lift off from the Moon with much of the equipment sent to the surface. Next, it will rendezvous with the first stage spacecraft, then all of it will return to the Gateway to be used on another traverse or a return trip to Earth.

The lander is to be a three-stage vehicle made up of a transfer vehicle, a descent module, and an ascent module. Some of the advantages cited for this system infrastructure are that it will be sustainable and offer international and commercial partners many opportunities.

The Gateway will also be a depot for propellant brought up to it after the propellant is generated from water ice on the Moon. In essence, it will be a storage facility, a haven, a spaceport for the crew, and a base for robotic equipment. Sound complex? Actual human exploration down on the lunar surface will begin no earlier than 2024 and, most likely, several or many years later.

A simplified statement of NASA's Gateway goals is "The United States will lead, using industry and international partnerships, as required, the development and demonstration of hardware and technologies for long-term human operations and science explorations on the Moon, Mars, and other deep-space destinations."

Even after all this is stated, it appears that NASA is about to deviate off a straight-line route to this goal. Simply put, yet painful to say, the Gateway is an unnecessary deviation, a waste of our current and future limited funding, time, talent, and opportunity!

These goals can be more simply and efficiently met by putting the Gateway on the lunar surface first. For our purposes, the Moon can be a much more effective and cost-efficient space station than one we manufacture and floats far above where we ultimately want to be. We cannot find a better place to put ourselves to study the Moon than on the Moon itself!

Additional objectives can be economically met by observations from Earth, Earth orbit, or one or more unmanned lunar orbiters. The author's view is that we should focus on near-continuous surface operations in which almost all the equipment, crew, and living quarters are first and permanently placed at one or more polar sites on the lunar surface. Initially, sites can be modest and then grow as required. Once on the surface, rocket propellant can be extracted from water ice or the lunar regolith to enable excursions to other lunar sites and return to Earth orbit.

Credit: NASA/Paul D. Lowman Jr.
Initial small lunar outpost

Strategically, a surface station will serve as a base for continuous explorations of the surrounding lunar surface, as a staging base for missions to Mars and farther out, and as a source of radiation shielding. There is no need to waste three-stage orbital equipment to transfer crews down and up or propellant to return to a gateway only to land again on the Moon. The crews can work on and explore the Moon continuously rather than in spurts of activity.

The surface infrastructure can be sustainable and offer international and commercial partners many opportunities. And it can also be a large storage facility, a haven, a space port for the crew, and a base for Being at a lunar pole and, with the Moon's one-sixth gravity, a hop from even a pole to a site on the equator will be a "short" hop in a lunar transfer vehicle and use much less propellant than a descent from an orbiting Gateway to a surface site, then a liftoff, rendezvous and docking with the Gateway to wait for the next opportunity to repeat the process.,

Let's put our home, office and lab down at surface site and work there or `

If we are going 250,000 miles to the Moon, let's skip the middleman and get down on the surface, explore it and utilize it! (Please see the Appendix for additional discussion of the Gateway.)

In the Longer Term

Regardless of the methods and routes taken, America will eventually return to the Moon, and it is likely that we will not be alone.

The drive to explore the lunar surface to define better the Moon's origin and future and those of our Earth and Solar System cannot be ignored indefinitely. Also, lunar explorations will technically, motivationally, and eventually lead expeditions to Mars and farther out. International prestige and leadership will continue to play a role. But in the long term, perhaps the biggest driver will be the economy. It remains to be seen if mining lunar minerals can be profitable. However, humanity's insatiable need for ever greater supplies of cheap, pollution-free energy will force us to take the potential of energy from the Moon's Helium 3 seriously—very seriously! Once harvesting this resource is proven to be economically feasible, the rush will be on.

As Harrison Schmitt has described in *Return to the Moon* (3), "A lunar surface station could become an economically independent lunar settlement, based on exports of consumables for other space operations and Helium-3 fuel for terrestrial fusion power plants as well as for interplanetary fusion rockets. Such fusion rockets may be necessary to shorten trips to Mars and other solar system destinations significantly."

Our stations on the lunar surface will grow in size and complexity. They will eventually include water extraction and propellant production capabilities, solar panels, small nuclear reactors if required, robotic and human-operated rovers, flyers, and equipment for electrolysis, refrigeration, and transportation of lunar materials.

Some early station concepts are shown on the next page. Those erected inside lava tubes are yet to be imaged.

Credit: Bigelow Aerospace
Inflatable lunar base by Bigelow Aerospace

Lunar Colony

Extraction of lunar minerals and ice are near-term objectives for space mining`

Our astronauts have explored the Moon six times since 1969. In 1972 Gene Cernan and Jack Schmitt were the last Americans to do so. After 1972 we've had the ready technology, a stable full of ready and eager astronauts, and a large and growing collection of scientific and economic motivations to continue our outward growth much farther away from Earth.

So, since then, since a half century ago, just where have we humans traveled beyond low Earth orbit?

NOWHERE!

After our early strong entry into space, our future in space appeared grim for many years.

Grade school kids who watched Neil Armstrong first land on the Moon are now retired.

Why?

There is a depressing multitude of economic and political excuses for this extreme delay:

1. After our default in 2011, we were unable to launch American astronauts from American soil. However, American free enterprise in the form of SpaceX just recently corrected this fault in 2020, and Blue Origin and/or Boeing are also ready to step in.
2. Our launch inadequacies had limited American participation in the ISS.
3. America's unmanned explorations were significantly restricted.
4. We had no feasible end-to-end program in place to return to the Moon and stay.
5. We had no firm plans to develop the technologies required to continue to Mars.
6. A sense of urgency to move quickly, efficiently, and cost-effectively did not exist throughout many of our major contributors.
7. America's eroding commitment to our space effort over nearly three decades is evident in the figure in Captor NASA's percent of our Federal Budget.
8. President Bush's staff lack of timely financial commitment to his Vision for Space Exploration caused the effort to slip and thereby come in over schedule and over cost, which teed it up for cancellation by the following Obama administration.

9. President Obama shut down America's ability to launch astronauts from American soil and saw no reason to return to the Moon because "we have been there before."
10. Critical flight-enabling hardware in the government pipeline is still not operational as of 2020 despite excessive investments of money (stated below in 2018 dollars) and time (14 years), which includes their corresponding developments in the Constellation Program (2005 to 2010). Thus, cost has become a prime issue.
11. Orion Spacecraft—$20 billion to develop and $1.0 billion per launch (not reusable)
12. Space Launch System—over $24 billion to develop and up to $4.1billion per launch.

The above cost numbers are difficult to pin down and are approximate only.

NASA, which had grown more bureaucratic and risk-averse, has set its course to return to the Moon through its Lunar Gateway. This unnecessary and inefficient deviation requires extra financial investment and time and will not be operational until at least 2028 unless scaled back in scope.

More generally, the focus and strength of our achievements in space often swing significantly depending upon the interests, knowledge, political leanings, and international challenges faced by each new presidential administration, which is often to the detriment of ongoing space efforts that require consistent and continuing support to achieve results and be cost-effective.

It's time to put all this failure to perform behind us.

It's time to again break out a big smile…

Positive change is here!

ARTEMIS

SLS rocket holding an Orion spacecraft payload preparing for launch.

Artemis is the new NASA program to explore the Moon. It will use innovative technologies to explore more of the lunar surface than ever before and land the first woman and first person of color on the Moon.

But where did the name Artemis come from? The name Apollo came from stories in Greek methodology many years ago. In these Greek stories, Apollo was the god who had a twin sister named Artemis who was the Goddess of the Moon.

NASA is collaborating with commercial and international partners, and to establish the first long-term presence on the Moon. Then, NASA will use what we learn on and around the Moon to take the next giant leap outward: *landing the first humans on Mars.*

In addition to dedication and drive, Bill Nelson has a rich and lengthy background. After he earned his law degree, he served 6 years in the Army Reserve (2 years on active duty), then 6 terms in the US the House of Representative and 4 terms in the US Senate.

Also, on January 12-18, 1986 he flew as a Payload Specialist on Space Shuttle Columbia's STS-61-C mission for 6 days. Ten days after landing, the Challenger tragedy occurred.

In May 2021 he became the NASA Administrator. As Administrator, has overseen the deployment of the James Webb Space Telescope, Artemis 1 and DART asteroid impact missions.

In addition to his drive to succeed, he is also a strong believer in *Safety Comes First*.

Oayload
NASA Administrator Bill Nelson

The leadership of NASA Administrator Bill Nelson is one of the major drivers that has kept Artemis on track.

Artemis – Let's get to Mars!

Artemis 1 was the first integrated flight test of NASA's deep space exploration system: the Orion spacecraft Space Launch System (SLS) rocket and the ground systems at Kennedy Space Center in Cape Canaveral, Florida. It is the first in a series of increasingly complex missions.

Artemis 1 was an uncrewed flight, except for "Commander Moonikin Campos," a mannequin outfitted with sensors to gather data on what crew members could experience on future flights. This mission has provided a foundation for human's next deep space exploration and demonstrated our commitment and capability to extend human existence to the Moon and beyond. Please see the Artemis I Trajectory in the Appendix.

During this flight, the Orion spacecraft launched on the most powerful rocket in the world and traveled thousands of miles beyond the Moon, farther than any spacecraft built for humans has ever flown, over the course of its 26-day mission.

Artemis 1 has traveled 270,000 miles from Earth, over the course of its four-week mission. Orion stayed in space longer than any ship intended for astronauts has done without docking to a space station and returned home faster and hotter than ever before.

"This is a mission that truly will do what hasn't been done and learn what isn't known," said Mike Sarafin, Artemis I mission manager at NASA Headquarters in Washington. He stated that it enables the next Orion flights by "pushing the edges of the envelope" to prepare for those missions.

What Spacecraft Will Be Used for the Artemis Program?

NASA has a new rocket. It is the Space Launch System, called SLS for short. It is the most powerful rocket in the `world. On top of the booster the Orion spacecraft will be positioned, which can carry up to four astronauts as it orbits the Moon and returns.

On future Artemis missions, a mini spaceship will be orbiting the Moon as the Moon orbits Earth. This spaceship is called Gateway. (Please see the discussion of the Gateway in the APPENDIX.) Orion will connect to Gateway, astronauts will enter it, and work and live in it. The crew will take trips in spacecraft called landers to get to work on the surface of the Moon. Then they will return to Gateway. When all of their work is finished, the crew will return to Earth aboard Orion.

Leaving Earth

SLS and Orion blasted off from Launch Complex 39B at NASA's modernized spaceport at Kennedy Space Center in Florida on November 16, 2022. The SLS rocket is designed for missions beyond low-Earth orbit carrying crew or cargo to the

Moon and beyond, and produce 8.8 million pounds of thrust during liftoff and ascent to lift a vehicle weighing nearly six million pounds to orbit. Propelled by a pair of five segment boosters and four RS-25 engines, the rocket hit its greatest atmospheric force within ninety seconds after launch. After jettisoning the boosters, service module panels, and launch abort system, the core stage engines shut down and the core stage separated from the spacecraft.

As the Orion spacecraft orbited Earth, it deployed its solar arrays and its Interim Cryogenic Propulsion Stage (ICPS) that gave it the big push needed to leave Earth's orbit and travel toward the Moon. Orion then separated from the ICPS about two hours after launch.

The ICPS deployed a number of small satellites, known as CubeSats, to perform several experiments and technology demonstrations.

On to the Moon

As Orion continued on its path from Earth orbit to the Moon, it was propelled by its service module provided by the European Space Agency, which supplied the spacecraft's main propulsion system and power. Orion passed through the Van Allen radiation belts, flew past the orbit of the Global Positioning System (GPS) satellite constellation and above communication satellites in Earth orbit.

The outbound trip to the Moon took several days and enabled engineers to evaluate the spacecraft's systems and correct its trajectory. Orion flew to about 62 miles above the surface of the Moon, then used the Moon's gravitational force to propel it into a retrograde orbit that revolves in an opposite direction to the Moon's revolution around Earth. The highest altitude (apogee) of the orbit was about 40,000 miles above the Moon.

The spacecraft stayed in this orbit for approximately six days to collect data and allow mission controllers to assess the performance of the spacecraft.

Return and Reentry

For its return trip to Earth, Orion did another close flyby of the Moon that took it within about 60 miles of the Moon's surface, where it fired the engine of the European-provided service module. At this low altitude and in conjunction with the Moon's gravity, the Orion accelerated back toward Earth.

On December 11, 2022, the Orion capsule separated from its Service Module and entered Earth's atmosphere at 25,000 mph, which produced temperatures of approximately 5,000 degrees Fahrenheit that is faster and hotter than Orion experienced during its 2014 flight test.

After 25 days and a total distance traveled exceeding 1.3 million miles, the mission ended with a test of Orion's ability to return safely to the Earth's surface. It landed off the coast of Baja, California.

Future Missions

With this first exploration mission completed, NASA is continuing to the next human explorations into deep space where astronauts will build and begin testing the systems near the Moon needed for lunar surface missions and exploration.

The time is now…

Let's explore!

Ed Gibson

Chapter 20
Moon First

Our Solar System – Inner Planets

MERCURY

Credit: MERCURY - Color Base Map from Messenger Mission

As seen in the figure above, Mercury, a rocky planet like Earth, is just slightly larger than our Moon with a diameter of 4,880 km. It has no moons, and its axis has the slightest tilt of any of the Solar System's planets (about 1/30 degree). Its sun is side roasted to a temperature of 427 °C while the side away from the Sun cools down to -173 °C.

Although small, Mercury is very dense because it consists of approximately 70% metallic and 30% silicate materials; its density is the second highest of all the planets except for that of Earth. Mercury's surface appears heavily cratered and is similar in appearance to that of our Moon, indicating that it has been geologically inactive for billions of years.

VENUS

Venus is slightly smaller than Earth at a diameter of 12,104 km. Though usually obscured by thick clouds, the overhead view was obtained from radar data acquired over a decade of satellite observations. It does have a small metal core, small magnetic fields, and no moons. Its atmosphere is 96.5% carbon dioxide, 3.5% nitrogen, and trace gases.

EARTH

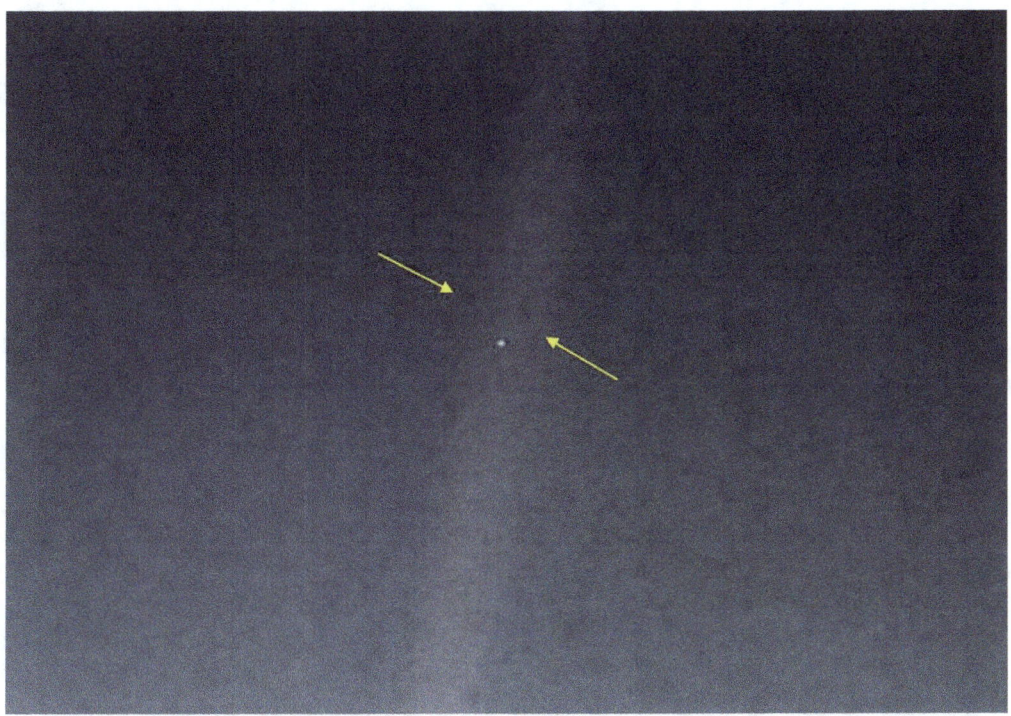

Just a pale blue dot

The above image of a pale blue dot between the orbits of Venus and Mars was taken by Voyager 1 in 1990 while near Saturn. But who could ever care about this tiny speck of dust except for any life that might happen to be living there?

Like US… We Humans… Here on Earth!

As much as life has become attached and dependent upon this little blue dot with its history and capabilities, Human realizes that there is much more unique yet livable turf out there. How about 1,000,000 times more? It could even be 1000,000,000 times more. Who knows?

We have a staggering number of highly different forms of life on Earth. What should we expect to find in an environment that is highly different than what we have on Earth? Likely, we are not even framing the questions correctly. If any, would life forms that we might encounter be larger or smaller, stronger or weaker, warm or cold-blooded (if they have blood at all), more or less intelligent, more or less aggressive, more or less technically advanced, and have a more or less intelligent approach to life. Some day we will begin to know, to understand and be able to do more than just wave our arms and guess (like what I am doing here).

Ed Gibson

MARS

Mars

Though larger` than Mercury, Mars is still the second-smallest planet in our solar system. Regardless of size, many generations have pinned their hopes for finding life in our solar system on Mars. Perhaps it once had life, but if so, is any of it still there today?

Four rovers have explored the surface of Mars: Sojourner, Opportunity, Spirit, and Curiosity. In June 2018, NASA announced that Curiosity had detected organic molecules in sedimentary rocks about three billion years old!

A fifth rover was launched to Mars at the end of July 2020 and landed about nine months later. Using its nuclear power supply for electricity generation, it can stay for at least one Martian year (687 Earth days). After landing in a crater that was once a lake 50 km wide, 0.5 km deep, and connected to a network of rivers over three billion years ago, it is in a prime location to search for evidence of past life. It will also be used to help us set up for the first human landing and exploration of Mars in the decade of the 2030s.

NASA Concept for a Future Rover and Station on Mars

Mars' surface gravity is 38% of our surface gravity on Earth. Hence, our weight would be 38% of what it is here on Earth. Unfortunately, for all of us, our mass will remain unchanged.

The mean surface temperature on Mars is a cool -81 °F. However, the surface temperature can reach 70 °F at the equator but remain as low as -250 °F at the poles.

In the past, its water was in lakes and rivers, but now, it has dried up or in the form of frozen dirt or thin clouds. Its thin gaseous atmosphere (0.006 of Earth's atmospheric pressure) is composed of primarily carbon dioxide (95.3%), small amounts of molecular nitrogen (2.6%) and argon (1.9%), and trace amounts of other gases (0.2%). Our Oxygen would have to be produced from carbon dioxide.

The dust in its atmosphere contains perchlorate, toxic compounds of chlorine. Samples of the Martian soil are thought to be contained in a few meteorites, rocks that ejected from the Martian surface and found their way to Earth. Their composition is the same as detected by our Martian Rovers. Mars has significant dust storms that can last from weeks to many months. We lost Curiosity to such a long storm. Because of the toxicity of the dust, special care will be required in handling any spacewalk equipment that comes in contact with it.

The composition of Mars is differentiated; that is, it has a dense core of iron and nickel surrounded by a mantel and a crust of silicates. There is no evidence of internal convection. Hence, without the motion of iron in its interior, Mars has no significant magnetic field.

Mars has two very small, irregularly shaped moons.

 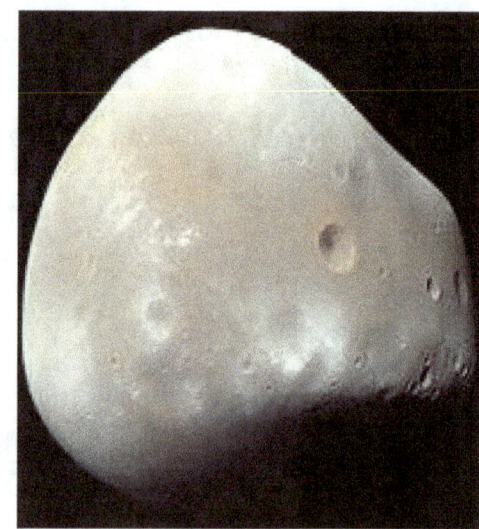

Credit: Calvin *Phobos* J. Hamilton *Deimos*

Curiosity has determined that life could have been supported by the Martian environment on the surface 3.5 billion years ago. There could still be some forms of elementary life deep in its soil. Curiosity has characterized the geology and climate of Mars and provided valuable information in preparation for human exploration. It survived two significant technical glitches, and it continues. InSight, a stationary research station, has joined the robotic explorers already on Mars.

And now a stationary observatory, Insight – 2018!

A selfie was taken by InSight on December 6, 2018

Credit: NASA/JPL Caltech (both pictures)

Sensing Improvements

Temperatures at InSight's location at the equatorial plain Elysium Planitia fluctuate by about 170 degrees Fahrenheit over the course of a typical Martian day. InSight's mission was scheduled to last at least two Earth years.

In March 2019, investigators were rewarded with the first detection of tiny Martian "baby" tremors. In July 2018, MARSIS, a radar instrument on the European Space Agency's Mars Express orbiter, announced the discovery of a persistent body of water on Mars — a lake that is a minimum of three to four feet deep, twelve miles long, and well under the southern polar ice cap.

The Martian surface currently does not support life as we know it, but it could exist in the subsurface where the lack of strong radiation and presence of liquid water might permit it. In parallel, it has been suggested that Mars volcanism is not completely dead, small amounts of subsurface magma could still exist, and its heat help retain some form of life.

`The detection of this lake could stimulate a new Mars mission that drills down into this water and analyzes it or returns samples. However, a precise rover landing and penetration of about a mile of ice with a robot drill would be required. Another location of past water, perhaps even a small amount of current underground water, is shown by the river valley in the picture below.

Ed Gibson

Credit: ESA/DLR/FU Berlin/CC BY-SA 3.0 IGO
Seeing Riverbeds of the Past on Mars!

The European Space Agency's Mars Express Orbiter has obtained images (above) of a dried-out river-valley network like we see on Earth. Very small then larger and larger downstream tributaries feed into the main river.

Preceding Opportunity, Spirit, Curiosity, and Insight, other American spacecraft have gone to Mars, including Viking 1 & 2 Orbiters and Landers (1975), Pathfinder Rover (1997), and Phoenix Rover (2007). There are also four active American Mars orbiters: Global` Surveyor, Odyssey, Reconnaissance, and MAVEN. Additional robotic explorers will come, each more capable and seeking more detailed and sophisticated information than those before. Eventually, humans will arrive.

CHALLENGES

Explorers and settlers moving into new territories on Earth have faced significant and continuous challenges; the unknown territory of space is proving to be no different. As just discussed, our return to the Moon presents new and very significant challenges. Mars will offer greater challenges than we have ever seen and ever more so for sites outside of our solar system.

Radiation

A significant challenge that humans will face as we start to explore Mars is radiation. On Earth, we have a strong magnetic field. In fact, we wouldn't be here without it. It not only diverts our compass needles to point north, but it also diverts high energy particles from the Sun so that almost all of them go around Earth rather than hitting us directly.

Earth's Magnetic Field Protects Us from Solar Radiation. Mars has not been so fortunate. It is smaller than Earth, cooled more quickly, and the motion of its iron core dropped to zero. Thus, its strong magnetic field faded away about three to four billion years ago. Its unprotected atmosphere was blown away, and its surface was then fully exposed to high-energy particles from the Sun and other cosmic sources.

Various levels of radiation that we can experience, including on a trip to Mars, are shown below on a vertical scale that increases by multiples of 10. Two levels are of prime interest. First is the US Annual Average, which is what we all experience down here (second dark blue horizontal bar). At the bottom is what we would experience on a trip to Mars, or a 500-day stay, each of which is about 100 times what we all experience (last two bars) now.

Even without additional solar radiation bursts, we will face significant radiation that will exceed NASA or self-imposed limits. Once on Mars, we can seek protection in lava tubes or other structures.

Exploration Radiation Threats

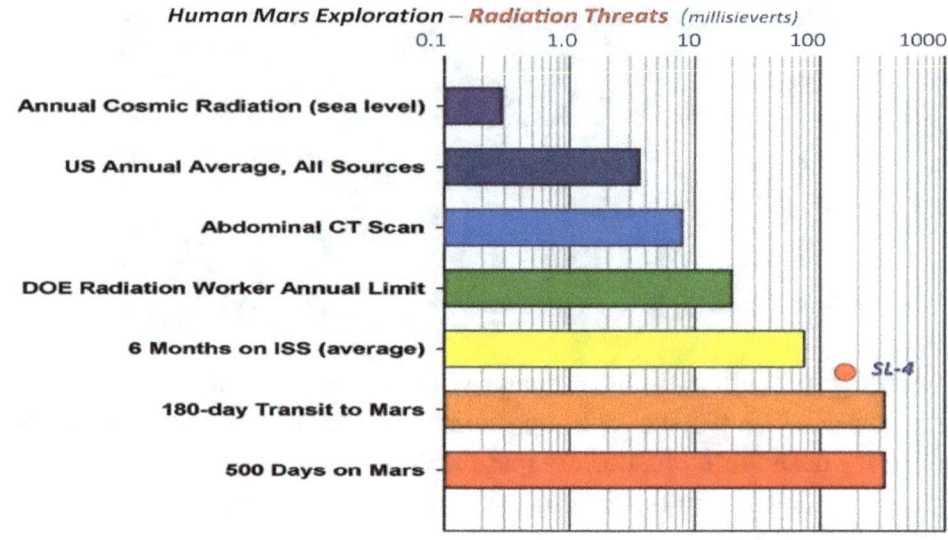

1 Sievert (100 rem) = Astronaut Career Limit

On our Skylab III mission, which was at an altitude of 270 miles and an orbital inclination of 50º, we experienced a higher cumulative radiation over 84 days than other crews to that time (red dot SL-4 in above plot). The radiation was exceptionally high while in the South Atlantic anomaly where we scraped along the bottom of the inner van Allen Radiation belt. We wondered if we might glow. When I got home, would I hear, "Hey Mom, Joey's afraid of the dark. Get Dad in here!"

Each of the above limits could be significantly exceeded if one or more strong solar flares or storms were to occur during a mission and emit high-energy particles towards a flight crew. Higher radiation levels increase cancer risk; female thresholds are about 20% less than that of males because of additional risks of breast, uterine, and ovarian cancers. Other potential risks, but so far ill-defined, arise from degradation of our immune system and short-term memory as well as possible mutations of microbes on our human skin surface.

Materials rich in Hydrogen, like water or Hydrogen propellant, are efficient radiation absorbers and can be designed to surround spacecraft crew quarters. The design of lightweight radiation absorbers is a high priority to enable future human missions to Mars and beyond. Under current technologies, the best advice is to go there and return quickly and live inside radiation-protected crew quarters while in transit and underground while on Mars.

Space, Ever Farther, Ever Faster – Now!

Credit: "The Martian" via 20th Century Fox

Gypsum is also present that, although not toxic, can build up in the lungs leading to a decrease in lung capacity like what coal miners on Earth can experience from coal dust. Since the dust is like a fine talcum powder, any breathing of the Martian atmosphere will cause these particles to be ingested. Compared to much of the dust that we experience here on Earth, dust on Mars is thought to be larger and rougher, approximately 3 to 30 microns in size, like the dust that covers the Moon. When astronauts ventured out on moonwalks, they soon found themselves covered by the rough, coarse lunar dust, which clung, scratched lenses, and degraded seals.

Credit: NASA/JOL   (Life*?*)

The Mars 2020 rover, an astrobiology mission, will make further measurements to help us understand the hazards of Martian dust in detail. So far, the Mars rovers have told us that within Mars, in addition to the hydrocarbons on its surface, Mars also once had water, Oxygen, and energy — the ingredients for life! `

Time of Travel

We have another challenge in just reaching Mars. We can't launch and return as we please but must wait for the opportunities as they arise. Because Mars takes approximately twice as long to go around the Sun as we do, it is only relatively close to us about once every 26 months.

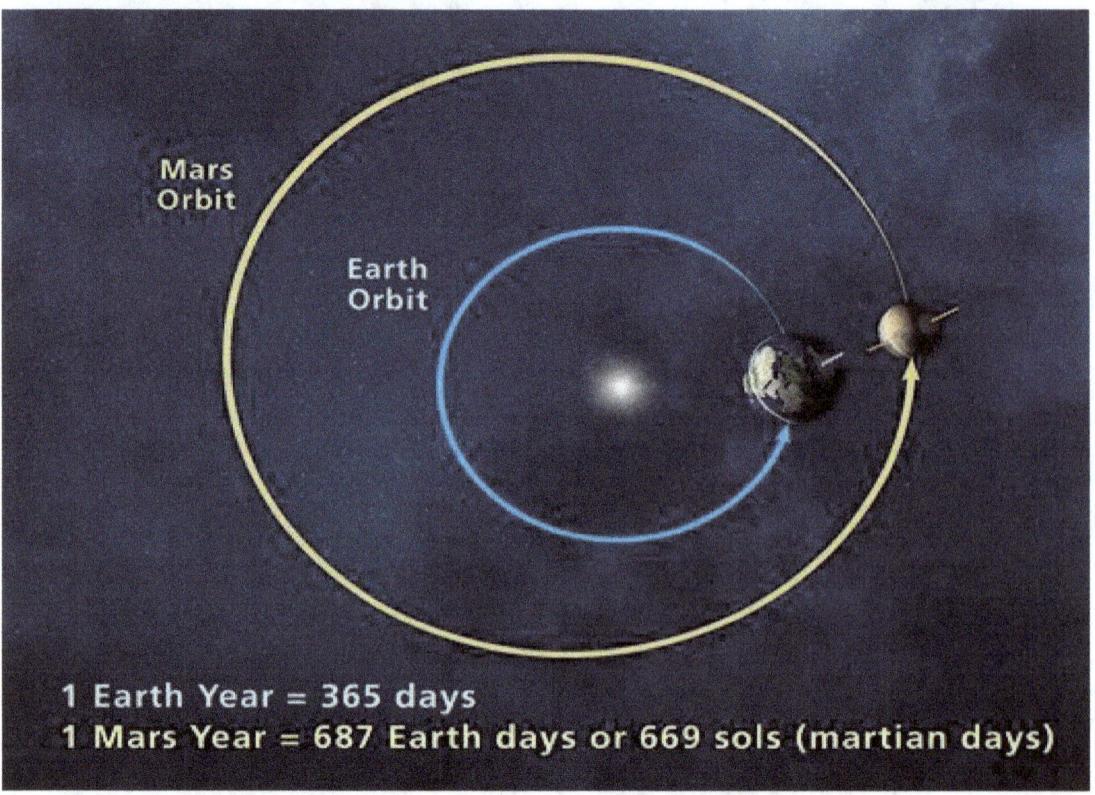

Mars takes almost twice as long to go around the sun as we do.

The energy needed for transfer between planetary orbits, or "ΔV," is lowest at intervals fixed by the synodic period, which is the time required for Mars to return to the same position relative to the Sun as seen from Earth. For the Earth-Mars system, this period is 780 days. The next low-energy launch window opens in late 2024. The lowest energy transfer to Mars is a Hohmann transfer orbit named after the German engineer Wolfgang Hohmann who first calculated it in 1925. It is an orbit that has rocket engine burns at the lowest point in the orbit (perigee) and the highest point (apogee).

Using Hohmann transfer, a mission to Mars involves approximately nine months of travel time from Earth to Mars. Moreover, about five hundred days at Mars to wait for the next lowest energy transfer window back to Earth, and a Hohmann transfer of again about nine months to return. A longer-time but lower-energy transfer is the ballistic capture transfer, which is being proposed and may be used for some future missions.

Once additional energy is expended to enter a Mars orbit, descent to the surface can be accomplished in part by using aerobraking in the Martian thin atmosphere to help reduce the energy required for landing.

Space Junk

All current and future space missions fly in an environment filled with ever-increasing amounts of "space junk." Old rocket boosters, current and now dead satellites, and debris from explosions and collisions are all hazards. America and Russia have each contributed a little over 4,000 pieces, while China has added about 3,500 pieces to the total of 12,000 pieces. Recently, early in 2017, India launched 104 satellites on its Polar Satellite.

A bb hitting one of us at several thousand miles per hour would undoubtedly shorten our day. In 133 Shuttle missions, orbital debris hit the spacecraft 1894 times. Fortunately, no fatalities have arisen from the hits encountered in space—so far. Unfortunately, we humans have no practical way to clean up all of humanity's space debris or even prevent it from growing.

It's only a matter of time until a fatal encounter occurs!

Credit: Wikipedia

In addition, as of 2022, we have started developing satellites to impact and divert asteroids that appear to be headed our way.

Direct Human Challenges

Other issues in our human colonization of Mars could arise from prolonged exposure to near-zero gravity on the way to and the return from Mars and the 38% of Earth's gravity on Mars. Eyesight problems in a few of the crew are to be expected (including cataracts and decreased visual acuity from eye structural changes), as well as issues of reduced strength and balance upon arrival. Also, being away from Earth for exceptionally long periods, limited communication with Earth and family, and living in close quarters continuously with other crew persons for two or three years could induce psychological issues.

Also, as has been done to our rovers and other spacecraft that have touched down on Mars, future spacecraft will be sterilized to prevent contamination of the Martian water or its environment. Humans and their spacesuits on Mars cannot be sterilized to the same rigorous standards as equipment, and concessions will have to be made.

China Competition

As seen previously, China has been active in Earth's orbit and on the Moon. Mars is also in its sights. It has been behind America and Russia in its space exploits since the beginning of the space race. The humiliation China has suffered has made it determined that this situation will not continue. (15)

China was late to the space race—it did not send its first satellite into space until 1970, just after the United States put the first men on the Moon. China dominated many mass-production markets during that period, although often with inferior products sold at 5&10 cent stores. Now they have shifted to high technology and innovation. Over the past 30 years, they have started down at the roots and made massive investments in science, technology, engineering, and math (STEM) education. Unfortunately, they have also indulged in technology theft, making nations and corporations that might work with them extremely cautious. But their efforts are beginning to bear fruit.

They have vowed that China, not America, will dominate technical innovation. It has made significant progress and has now become a space power, having sent five crews into space, making it only the third country in the world, after Russia and the United States, to do so. They have performed spacewalks, landed a lunar rover, launched a space lab in preparation for a much larger space station, but have not landed on Mars.

Credit: Xinhua

Artist Impression of China's 2020 Mars Lander and Rover

The launch of China's Mars Lander and Rover occurred in July or of 2020 during the once-every-26-months minimum-energy Mars launch window. We still await a detailed report.

Technically, China has advanced in many fields over the past 70 years or so. Honest competition is good; stealing technology and increasing their military strength far beyond what is required for their own protection is not. They must be carefully watched and dominated in technical areas that threaten our own security.

On July 23, 2020, China launched their Xinhau Mars Lander and Zhurong 530-pound rover that was deployed on Mars on May 22, 2021. It appeared to be similar to the vehicle that they previously had deployed on the Moon. It had made a 9-minute-long reentry before making a soft landing on the Martian surface. China was the only nation other than the United States that had landed a fully functional spacecraft on the Martian surface. It studied the surface using ground-penetrating radar, magnetometer, microphone, laser-induced material breakdown and spectrometer to define the surface geology, soil, mineral and ice composition and disturbance and magnetic field structure. By July 11, 2021, Zhurong had traveled 1,350 feet and as of August 18, 2021 outlived its lifetime, yet continued to explore.

Clearly, in terms of exploration and technical capabilities, China is now on par with the United States.

But soon American footprints will be on Mars!

> Yep, mine. I'll use my lunar suit!

In our quick look at the inner planets, it still appears that if we consider another place to take up residence other than Earth, Mars and our Moon are our two prime possibilities. Led by NASA through the Artemis Program, there is a growing interest and support for human spaceflight across America today. An ardent core has our sights stubbornly set on a return to the Moon, then on a mission to beautiful Mars.

Eventually, once our Moon then Mars become familiar turf, we will naturally shift our sights farther out. But it's got to be much colder as we move farther out, frozen dead cold. We live in the Goldilocks' zone of our solar system: not too hot, not too cold, but just right for life.

Yet when we look farther out, we still seek past life, or even current life. It's in our human nature to wonder, to seek.

Space, Ever Farther, Ever Faster – Now!

Chapter 21
Our Solar System – Outer Planets

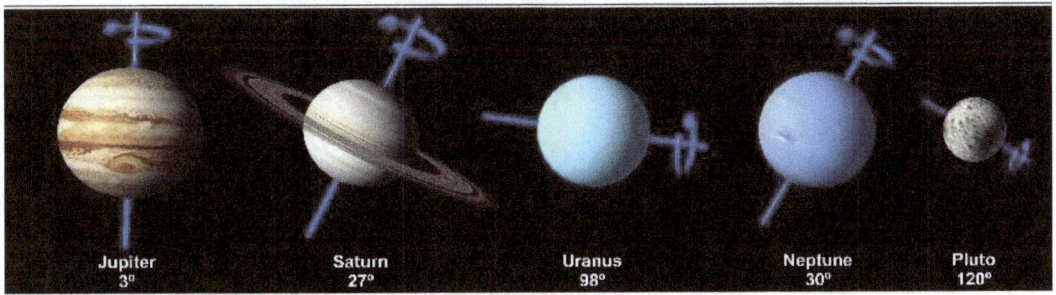

First, JUPITER - It's BIG!

GIANT RED SPOT

Credit: NASA Hubble Space Telescope

Credit: NASA SwRI/MSSS/Seán Doran

Giant Red Spot

This spot is thought to be a positive pressure storm that that is long-lived because it does not contact a solid surface underneath.

Jupiter has strong magnetic fields, which are responsible for the strong, variable radio emissions from polar regions. They also form its strong, far-reaching magnetosphere that protect many of its moons from the solar wind.

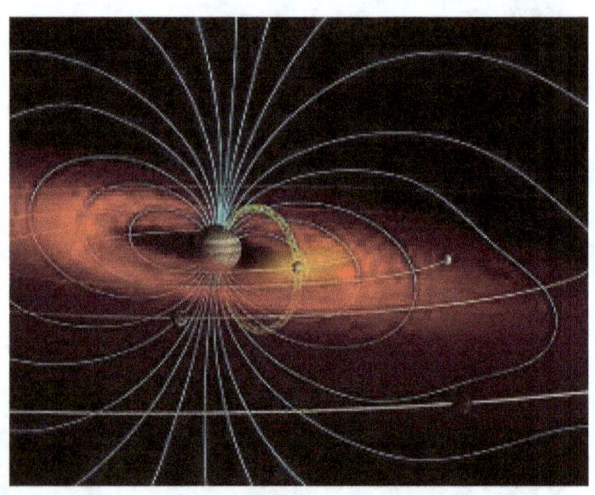

Credit: John Spencer's Astronomical Visualizations

Jupiter's equator is tilted with respect to its orbital path around the Sun by just 3 degrees. Thus, Jupiter spins nearly upright and does not have seasonal extremes like Earth and most other planets.

It is a gas giant with a mass of about one-thousandth that of the Sun, but two-and-a-half times that of all the other planets in the Solar System combined!

The composition of Jupiter is like that of the Sun—mostly Hydrogen and Helium, the same ingredients in a star. However, Jupiter never was massive enough to ignite.

Deep in Jupiter's atmosphere, its pressure and temperature greatly increase, compressing the Hydrogen gas into a liquid. This gives Jupiter the largest ocean in the solar system—but an ocean made of liquid Hydrogen, not water.

Some scientists speculate that, at depths perhaps halfway to the planet's center, the pressure becomes so great that electrons are squeezed off the Hydrogen ions (protons), making the liquid electrically conducting like a metal. It is believed that Jupiter's fast rotation drives electrical currents in this region, generating the planet's powerful magnetic fields. But it is still not understood if deeper down Jupiter has a central core of solid material or if it may be just a thick, super-hot, dense soup. It could be as hot as 50,000 degrees Celsius and mainly made of iron and silicate minerals like quartz.

Atmosphere

Jupiter's appearance is that of a colorful array of spots and bands. It likely has three distinct cloud layers in its "sky" that, taken together, span a height of about 44 miles (71 kilometers). The top cloud is probably made of ammonia ice, while the middle layer is likely made of ammonium hydrosulfide crystals. The innermost layer may be made of water, ice, and vapor. But all this speculation since we do not any have direct measurements.

Jupiter's magnetic field also causes some of the solar system's most spectacular aurorae at the planet's poles. (Please see the "It's BIG" picture of Jupiter at the beginning of this section.) Jupiter's prevailing winds can reach up to 330 miles per hour at the equator. With no close solid surface to slow them down, surface structures like its Red Spot can remain essentially unchanged for many years.

Dozens of moons surround Jupiter. It also has several rings, but unlike the famous rings of Saturn, Jupiter's rings are dark, very faint, and made of dust, not ice. Nine spacecraft have studied Jupiter up close. NASA's Juno spacecraft is currently studying the gas giant planet from its orbit. This spacecraft, which arrived at Jupiter in July 2016, helps us learn more about the planet's mysterious interior. Planetary scientists also use the Earth-orbiting Hubble Space Telescope and ground-based telescopes to check in on Jupiter regularly.

Pioneer 10 was the first spacecraft to fly past Jupiter. It was followed by the Pioneer 11, then the Voyager 1 and Voyager 2 flybys. NASA's Galileo mission was first to orbit Jupiter and send an atmospheric probe into its stormy clouds. The international Ulysses mission used Jupiter's powerful gravity to hurl itself into the Sun's northern and southern poles orbital passes. Both Cassini and New Horizons studied Jupiter as they hurtled on to their main science targets — Saturn for Cassini and Pluto and the Kuiper Belt for New Horizons.

Two new missions are in the works to make closer studies of Jupiter's moons: NASA's Europa Clipper and ESA's JUpiter ICy Moons Explorer (JUICE).

Jupiter holds a unique place in the history of space exploration. In 1610, astronomer Galileo Galilei used an invention called a telescope to look at Jupiter and discovered the first moons known to exist beyond Earth. The discovery ended incorrect, ancient belief that everything, including the Sun and other planets, orbited the Earth.

Orbit and Rotation

Jupiter has the shortest day in the solar system. One day on Jupiter takes only about 10 hours (when it takes for Jupiter to rotate or spin around once). In contrast, Jupiter makes a complete orbit around the Sun (a year in Jovian time) in about 12 Earth years (4,333 Earth days).

Rings

Discovered in 1979 by NASA's Voyager 1 spacecraft, Jupiter's rings were a surprise, as they are composed of small, dark particles and are difficult to see except when backlit by the Sun. Data from the Galileo spacecraft indicate that Jupiter's ring system may be formed by dust kicked up as interplanetary meteoroids smashed into the giant planet's small innermost moons.

As a gas giant, Jupiter does not have an actual surface, and the planet is mostly swirling gases and liquids. While a spacecraft would have nowhere to land on Jupiter, it would not be able to fly through unscathed either. The extreme pressures and temperatures deep inside would crush, melt, and vaporize any spacecraft trying to fly into it.

Magnetosphere

The Jovian magnetosphere is the region of space influenced by Jupiter's powerful magnetic field. It balloons out 600,000 to 2 million miles (1 to 3 million kilometers) toward the Sun (seven to 21 times the diameter of Jupiter itself). It tapers into a tadpole-shaped tail extending more than six hundred million miles (1 billion kilometers) behind Jupiter, as far as Saturn's orbit. Jupiter's enormous magnetic field is 16 to 54 times as powerful as that on Earth, and it rotates with the planet and sweeps up particles with an electric charge. Near the planet, the magnetic field traps swarms of charged particles and accelerates them to extremely high energies, creating intense radiation that bombards the innermost moons and can damage spacecraft. Jupiter's magnetic field also causes some of the solar system's most spectacular aurorae at the planet's poles.

Discovered in 1979 by NASA's Voyager 1 spacecraft, Jupiter's rings came as a surprise because they are made up of small, dark particles and are difficult to see except when backlit by the Sun. It is suspected that these dark particles resulted from collisions between Jupiter's inner moons other bodies such as meteoroids.

Could we, or would we ever want to try to live on Jupiter?

I agree. No!

Then . . . how about out further out? Moons maybe?

Jupiter's Moons	Diameter Compared to Earth's Moon
Io	1.05
Europa	0.9
Ganymede	1.5
Callisto	1.4

IO

In our Solar System, IO is the highest density moon. It has the most geologic activity and the lowest amount of water. Because of its proximity to Jupiter and the other large moons, it experiences large tidal forces and heating, leading to its high volcanic activity. Its frigid surface, at approximately -160°C, is coated with yellow regions of sulfur and more whitish plains of volcanically deposited sulfur dioxide frost.

Also, Io's surface has many mountains, some taller than our Mt. Everest. Its volcanism ejects large amounts of particles and gaseous plasma, which creates a patchy, thin, layer that and enters Jupiter's magnetosphere to form a torus of particles and plasma around the planet.

IO is the most volcanically active body in the solar system.

Ed Gibson

EUROPA

Europa is the sixth-closest Moon to Jupiter of its 79 known moons and the sixth-largest moon in the Solar System. Like IO, it has a cool surface of about -170 °C and surface gravity of just 1.4% of ours.

Credit: Wikipedia, The free Encyclopedia

EUROPA

Europa is made mostly of silicate rock with an Iron-Nickel core covered by about a 100 km thick layer of water-ice. Its fragile atmosphere is composed primarily of Oxygen. Europa has very few mountains and craters, making it the smoothest surface of any known solid object in the Solar System, but cracks and streaks cover it.

Could Europa also be a potential site for extraterrestrial life? With a surface radiation level of 540 Rems per day, it is unlikely to exist there. However, let's look deeper underneath Europa's icy crust where it is thought to have a huge, planet-wide liquid water ocean. All life on Earth requires liquid water, so what better place to look for life on Europa than within its ocean, where life would be protected from Jupiter's ambient radiation? Furthermore, its ocean would be relatively warm, which is not much below freezing temperature compared with the chilly −160 °C on the planet's surface itself. It is believed that its ocean is heated by Europa's tidal flexing and, locally, by erupting, hydrothermal ocean floor vents. Thus, its striated surface features see most likely result from shifting ice-plate tectonics below. These waters are one of the more possible places to find life elsewhere in our solar system.

Like other Galilean moons, Europa is tidally locked with Jupiter, almost. It spins just a bit faster than it revolves around Jupiter, providing it with another source of tidal heating. The Hubble Space Telescope has seen traces of water plumes ejected from its surface like those ejected from Enceladus, a Moon of Saturn, where it has been speculated that life could exist.

Let's look at another Moon.

GANYMEDE

With its diameter of 5,268 km, Ganymede is the largest and most massive Moon in the Solar System, even 8% larger than the planet Mercury, twice as massive as our Moon, and the largest solar system object without a substantial atmosphere. It is believed to be made of roughly equal amounts of rock and ice. Other than a very thin atomic Oxygen atmosphere, it has no breathable air. Yet, Ganymede is unique—it is the only moon in our solar system to have its very own magnetic field and associated auroras, probably induced by Jupiter's magnetic field.

Using the Hubble Telescope, scientists studied how those auroras changed with shifts in Jupiter's magnetic field and speculated that Ganymede has a vast ocean of salty water beneath its icy surface, so large in fact, that it could contain more saltwater than in all of Earth's oceans combined! Another idea is that Ganymede's interior is made from layers of salty liquid ocean interspersed with layers of ice, which results from how salted water and ice behave under extreme pressure.

What about life? Perhaps there is some life in its vast liquid ocean. After all, life developed within Earth's oceans. Why not Ganymede's? But could there be any form of life on the surface of Ganymede? There is some surface radiation like Europa, but it is much less at only 5-8 Rems per day.

This radiation level is still lethal, and you would surely die, but more slowly than on Europa!

Space, Ever Farther, Ever Faster – Now!

Credit: NASA

GANYMEDE

CALLISTO

Callisto has a surface temperature of about -140 °C. As evidenced in the image below, its surface is also the most heavily cratered Moon of Jupiter. Like other Gallian moons of Jupiter, it is tidally locked to its planet as it revolves. Its day and year are both about 16.7 Earth days, and its tidal heating is extremely low.

Callisto lacks any appreciable volcanic activity. Thus, its surface impact features were not erased except by other impacts. One huge and long-lasting impact feature, Valhalla, has a 3,800 km diameter Is shown to the upper right in the image on the precious page and in the image below:

Credit: Wikipedia, The free Encyclopedia

CALLISTO, Jupiter's most distant Moon imaged by Galileo Spacecraft

The next such feature on Callisto, Asgard, is only about half as large.

Callisto is the 2nd largest Moon of Jupiter and the 3rd largest in the Solar System. It is the lowest density of all of Jupiter's moons and has a surface temperature of approximately -140 °C. As evidenced by the first image in this section (above), its surface is most heavily cratered of Jupiter's moons. And, like other moons, it is tidally locked to its planet as it revolves. That is, its day and year are both about 16.7 Earth days.

Just like with Europa and Ganymede, as well as Saturn's moons Enceladus, Dione, Titan, and Neptune's Moon Triton, it is speculated that the subsurface salty ocean of Castillo could contain life. In the very salty waters on Earth, such as in the Great Salt Lake (UT), Owens Lake (CA), and the Dead Sea (between Jordan and Israel and at 1300 feet below sea level), which are at least five more times salty than Earth's other oceans, living organisms called Halophiles are found.

Valhalla creater on Callisto moon of Jupiter

Because of Callisto's low radiation at its surface, the low velocity required to escape the Moon's surface, and the full moon itself, Callisto is considered a good candidate on which to set up a base for human exploration of the whole Jupiter system as well as the production of rocket fuel.

Of the Galilean moons, Callisto is the furthest one from Jupiter. Hence, it experiences low life-quenching radiation, low gravity and gravity gradients from its mother planet, and low internal heating or tectonic activity.

Because of its large orbit that is outside Jupiter's magnetosphere and main radiation belt, the level of radiation at its surface is only about 0.01 rem per day.

With all this said, the environmental conditions necessary for life to appear might be even more favorable on Europa because its oceans are in contact with rocky material. It has a higher heat flux from its interior.

We do not yet know enough to pick the best conditions for life to flourish in our Solar System, other than Earth. Most likely, the final verdict will not be delivered until we explore each site up close and personal.

Much to learn! Much to do! Let us get on with it!

SATURN — The Other Gas Giant

Credit: Wikipedia, the free encyclopedia
*Saturn photographed in natural color by the
Cassini spacecraft in July 2008*

Saturn is the second-largest planet right after Jupiter and is also a gas giant. It is 84.4% the size of Jupiter but still 9.45 times the size of Earth. Although Jupiter is 3.5 times more massive than Saturn, these two planets make up 92% of all the mass, excluding the Sun in our Solar System! For a planet, it has a very low density of only 68% that of water. A few astronomers have observed that "Saturn would float in water if you could find a bathtub big enough."

Saturn's day is only 56% of ours. For a large planet, this rotation rate is amazingly fast. Thus, being mostly gas, its equatorial diameter is 110% of its polar diameter. This bulge and the planet's high rotation rate make the surface gravity at the equator only 74% of that at the poles. However, because of Saturn's large mass, the escape velocity at the equator is still large, almost 4 ½ times that at Earth's equator.

The tilt of its axis of rotation relative to the plane of its revolution around the sun is 26.3°, which is almost the same as Earth's average value of 23.3°. If seasons had meaning on a gas giant, they would be slightly more exaggerated than ours but more stretched out because it takes Saturn 29 Earth-years to go around the Sun just once.

Although called a gas giant, most of Saturn's mass is inside the lower boundary of its gaseous atmosphere. Deep inside the planet, there is a hardcore of rock and iron-nickel. Then a thick layer above electrically conducting metallic Hydrogen, a layer of liquid Hydrogen and liquid Helium. Finally, its gaseous atmosphere is mostly Hydrogen (96%), Helium (3%), and trace amounts of Hydrogen-containing compounds. Like all outer planets, the surface of Saturn is icy cold, with an average temperature of about −140 °C.

Space, Ever Farther, Ever Faster – Now!

The Rings of Saturn

The rings extend from 6,630 to 120,700 kilometers (4,120 to 75,000 mi) outward from Saturn's equator and average approximately 20 meters (66 ft) in thickness. They are made up of water, ice, dust, and trace amounts of impurities and range in size from dust up to approximately 10 m.

Despite the beauty of its rings, Saturn's surface is a pale dull yellow (above) because of the ammonia crystals in its upper atmosphere. The electrical current running through its inner metallic Hydrogen layer creates its outer magnetic field, which is relatively weaker. However, not bland nor weak are its surface winds that reach up to 1,100 mph.

Saturn's trademark is its beautiful, graceful system of rings. Although other planets also have rings, they are nowhere near as prominent and beautiful as those around Saturn. Saturn's core temperature reaches 11,700 °C, and the planet radiates 2.5 times MORE energy into space than it receives from the Sun, which does not come from nuclear reactions like in a star. So then, where does it come from?

It is generated by something called the "Kelvin–Helmholtz mechanism," which is also active in other gas giants and stars. As the top layer of the gaseous atmosphere emits radiation, it cools and contracts inward. The gravitational energy given up goes right back into the atmosphere below, which is now ever so slightly smaller as it becomes ever so slightly hotter. The small amount of heating is also increased slightly by small drops of liquid Helium that have condensed out in the cooling of the gas above, which then rain down through the gaseous Hydrogen below and frictionally heat it.

However, this increase in temperature is not free. Eventually, only a much smaller and colder ball of gas would be left if one could wait long enough as the atmosphere radiated energy away. It just takes a more circuitous route than anticipated to get there.

In addition to its rings, Saturn has another interesting visual feature, a hexagonal cloud pattern at its north pole. This rare cloud pattern was first observed by NASA's Voyager flybys of Saturn in 1980 and 1981. Many years later, close-up data from NASA's Cassini spacecraft confirmed its existence and provided details on its geometry.

This cloud pattern rotates simultaneously as the underlying planet's surface. At its center is a churning storm that is about 20,000 miles wide. Thermal data indicate that this storm reaches down into Saturn's atmosphere about 60 miles.

Following the outline of the hexagonal pattern closely, a jet stream flows eastward at 220 mph, a bit like the jet streams. This cloud pattern rotates at the same rate as the underlying planet's surface. At its center is a churning storm that is about 20,000 miles wide.

Hexagonal Cloud Pattern at Saturn's North Pole

The near-perfect hexagonal nature of this cloud pattern was not at all expected nor well understood. Its steady existence is a good indicator of a standing wave pattern and the lack of a hard surface underneath. A team headed by Raúl Morales-Juberías, a planetary scientist at the New Mexico Institute of Mining and Technology, has a model that matches many of its observed properties.

As we examine Saturn's interesting features, we must reluctantly conclude that this planet is also not a good prospect to find life nor support life that would be taken to it.

Moons of Saturn

Of Saturn's 82 known moons, only 53 are officially named. This count does not include the hundreds of tiny moonlets in its rings. Below some of the physical characteristics of Saturn's seven major moons are displayed as a percent of the corresponding attributes of our Earth's Moon. Note that Lapetus is three times further away from Saturn than any of its other significant moons.

Physical Characteristics of Saturn's Major Moons as a Percent of Our Moon				
Moon	Diameter	Mass	Orbital Radius	Orbital Period
Titan	148	180	318	60
Rhea	44	3.0	137	20
Lapetus	42	2.5	926	290
Dione	32	1.5	98	10
Tethys	30	0.8	77	7
Enceladus	14	0.2	62	5
Mimas	12	0.05	48	3

TITAN

Credit: NASA/JPL-Caltech/Univ. Nantes/Univ. of Arizona
*Titan Viewed in One Visible Light Image and
Five Ultraviolet Light Images*

Many aerosols in Titan's atmosphere reflect light providing its opaque yellow appearance. The five UV images above are made using five different narrow bands of UV in the near visible spectrum.

These pictures started with the data collected during the 13 years of observation by the Visual and Infrared Mapping Spectrometer (VIMS) instrument onboard NASA's Cassini spacecraft, which contained images from many different flybys with different observing geometries, lighting, and atmospheric conditions as well.

The objective was to produce pictures that best represented how Titan might appear if it weren't for its hazy atmosphere. It is likely that these resulting images will not be superseded for some time to come.

It is clear from these images that Titan has a complex surface with many different large and small geologic features. The VIMS instrument has paved the way for future infrared instruments that could image Titan at much higher resolution and reveal features that were not detectable by any of Cassini's instruments.

Titan's surface is dotted with many "beautiful" lakes, almost like resort areas in America!

Ed Gibson

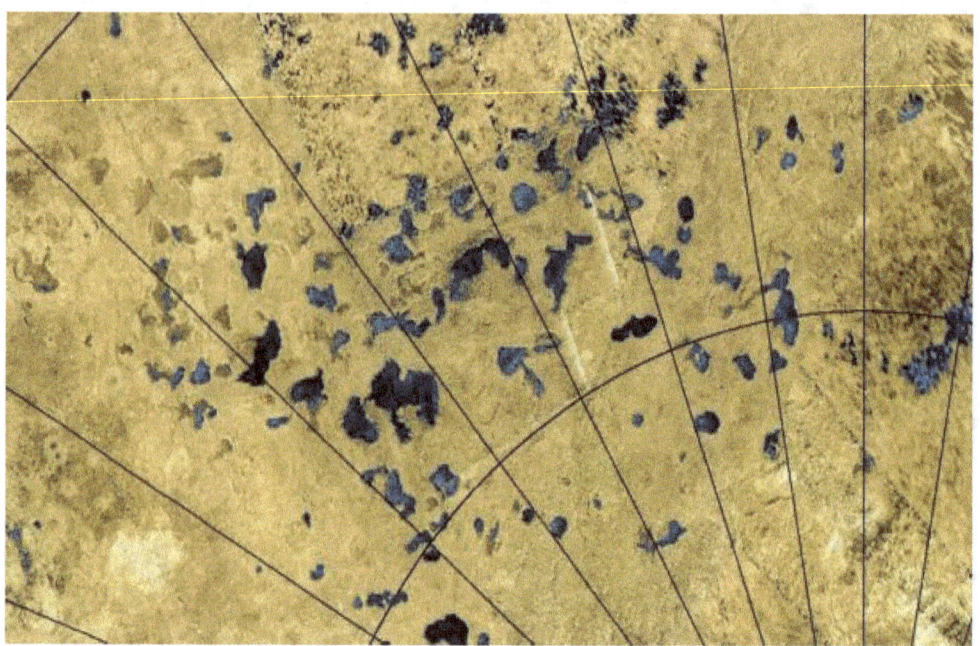

Credit: NASA/JPL-Caltech/ASI/USGS

Radar image of the lakes near Titan's North Pole.

But before we invest, let's read the fine print. It is believed that these lakes are mostly liquid hydrocarbons that fill depressions in the moon's soil. It is freezing like the other outer planets: -180 °C (- 292 °F). The lakes and rivers are liquid methane and ethane. If any life forms exist there because of the abundance of hydrocarbons, they would not be like anything we have experienced.

Another real estate non-selling point is location, location. On average, Titan is about 891 million miles away from Earth. Even traveling at the speed of light, it would take one hour and 20 minutes to reach Titan. Traveling at the more "reasonable" speed of just 67,000 mph (escape velocity from Earth is 25,000 mph), it would take 554 days (1.5 years) to reach it.

The only man-made object that has explored Titan face-to-face has been NASA's Huygens Cassini probe, whose mission ended in 2017. But when humans are on board the spacecraft, it is not just "travel time" that is important, but it's "the time spent in Saturn's strong radiation environment" that could bring mission planning to a halt.

As for wind power, Titan does have strong winds in its upper atmosphere, according to measurements made by the Huygens probe. However, currently, we do not have the technology to float efficient wind turbines at the required heights. This situation identifies the need to improve earth-bound technologies that could apply to space exploration, including much higher efficiency turbines and solar panels. Also, on Titan, there are generally abundant hydrocarbon fuels but the Oxygen to combust them is rare. This requires more investigation of where there might be other sources of Oxygen on Titan that we have not recognized so far.

Saturn's satellite system is very lopsided: one enormous moon, Titan. The six other ellipsoidal moons constitute roughly 4% of the mass, and the remaining 75 small moons, together with the rings, comprise only 0.0

In addition to the relatively massive Titan, could any of the smaller and more distant moons have any features of interest? When we get to Lapetus, we encounter the Equatorial ridge shown below.

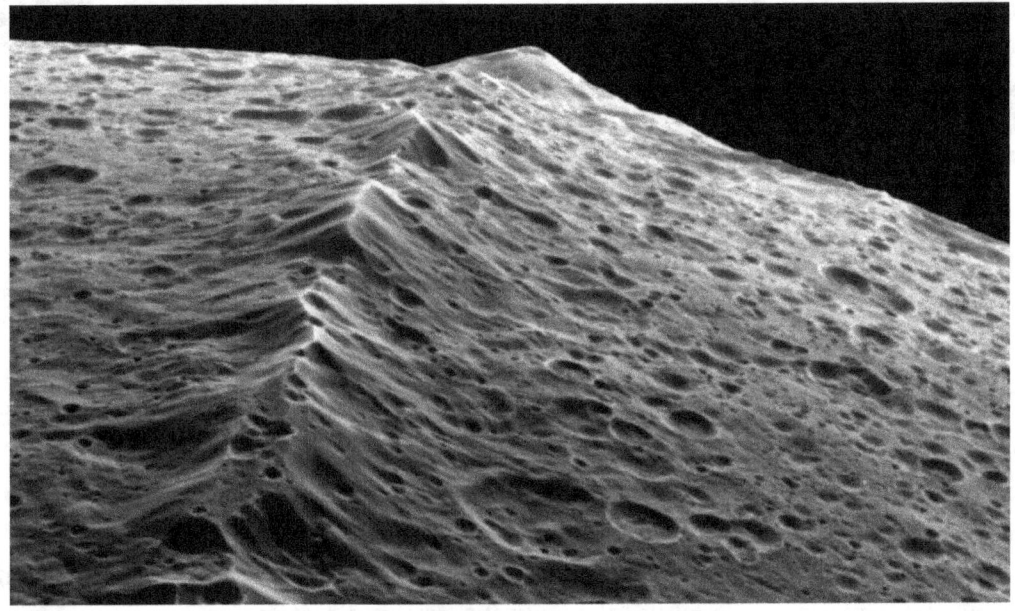

Credit: NASA/JPL

Lapetus Equatorial Ridge

Yes, Titan and Lapetus have some interesting features, but what about Life?

Those remaining and smaller moons outside the Goldilocks zone must be frozen solid cold. We shouldn't expect to find any life there. Nonetheless, the Cassini spacecraft was sent to look. When arriving at Enceladus, which is only 20% the size of our moon, a monumental surprise was encountered. Please see the image below.

Credit: NASA/JPL-Caltech

The south pole of Enceladus as seen from the highly successful Cassini mission

What? No ice? NO! Just jets of WATER vapor! Time for us to think again…

As Enceladus rotates in the gravity field of Saturn, the part of this moon that is closest to Saturn is pulled on with more force than part of the moon that is farther away. Thus, the moon is slightly stretched. As the moon rotates, the direction of this stretch rotates within the moon. This stretching and compression of every part of the moon heat it, and its ice melts. Sometimes the term "gravity gradient heating" is applied to this phenomenon. Once again, making assumptions about what we only partially understand can quickly lead us in the wrong direction.

Many investigators believe that if life is found elsewhere within our solar system other than Earth, it will be on Enceladus.

The highly successful Cassini-Huygens mission was a cooperative project of NASA, the European Space Agency, and the Italian Space Agency. NASA'S Jet Propulsion Laboratory, a division of Caltech in Pasadena, managed the mission for NASA's Science Mission Directorate, Washington. The VIMS team was based at the University of Arizona in Tucson.

ICE GIANTS

Uranus and Neptune are similar in composition to one another and the Gas Giants Jupiter and Saturn because of their presence of Hydrogen and Helium.

URANUS

When Voyager 2 arrived at Uranus on January 24, 1986, it was greeted by a giant cloud-covered planet, one of two such ice giants in our solar system.

Speculation implies that about the inner two-thirds of the planet is molten rock, liquid ammonia, and Methane. It does have a magnetic field that is highly tilted at about 47°, which is created within a conductive material that is most likely water. Its outer third probably consists of gasses heated by the planet's core with an icy mantle rather than a hard crust.

Uranus and Neptune are similar in composition to one another and the Gas Giants Jupiter and Saturn in their presence of Hydrogen and Helium. Still, they differ from the compositions of these gas giants, and they have higher percentages of these gases.

For this reason and their lower temperatures, they are termed "Ice Giants."

Credit: Solarsystem.NASA.gov. (NASA/JPL-Caltech)

Uranus has a magnetic field that is about the same strength as our field on Earth; however, the trail it leaves as it spins along on its side with its magnetic field axis about 60° offset from its spin axis is quite different.

In fact, Uranus leaves a long trailing magnetic corkscrew behind itself. Also surprising is that the planet's exposed cloud tops are almost the same temperature as its dark pole.

The haze at its sunlit pole radiates away a large amount of ultraviolet light termed an "electro glow." As expected, the planet's temperature is very low at 59 °K (−214 °C), which is just a shade above Neptune's surface temperature.

NEPTUNE

Neptune has earned the title of "The Windiest Planet in the Solar System" because winds there reach speeds of 2,000 km per hour (1,200 mph). The strongest winds recorded on Earth have been only 372 km per hour (231 mph), just 20% of the wind speeds on Neptune.

What drives these winds and the planet's heating? Unlike the Moon of Saturn, Enceladus, there are no nearby large sources of gravity gradients to stretch and compress the body as it rotates.

With negligible heating from the Sun and no nearby sources of gravity gradient forces on Neptune, the obvious question arises: Where do its high internal temperatures originate? Hmmm… needs some thought. The only answer left is residual heat left from its energetic formation. Neptune's large mass has provided layers of insulation that have significantly slowed the escape of its initial internal heat.

Credit: NASA Voyager 2 *Great Neptune's Dark Spot*

Neptune's atmosphere contains two layers. The gas in the upper atmosphere contains high levels of Hydrogen (80%) and Helium (19%) with traces of methane like other gas giants in our Solar system. Temperatures within this layer average about -218°C (-350°F), making this the coldest primary planet within our Solar System. In the second layer, which also consists of Hydrogen and Helium like the upper layer, the gasses are superheated to 7,000°C (12,632°F).

Periodically, large dark spots appear that can be even larger than Earth. They resemble storms as we know them on Earth, except that they are high-pressure areas, not low-pressure ones. Whether high or low pressure, their direction of rotation is driven by Coriolis forces like every location in the universe. High pressures areas on a planet or star will rotate clockwise, just like high-pressure areas on Earth. High-pressure areas are also areas of low humidity; and, therefore, low cloud presence and visibility.

Voyager 2 discovered that the dark spots were, in fact, "holes" within the atmosphere, which created windows through the cloud cover to view deeper within the planet. However, no other missions are planned to see what those holes might reveal.

Lastly, Neptune has 14 moons, with the most massive Moon Triton. Triton is the only large satellite in the Solar System that orbits its parent in a retrograde or backward motion.

Other small moons might orbit in this manner, especially if they are very distant from their planet. There is speculation that Triton was once captured from the farther-out Kuiper Belt by Neptune's gravity and now remains as one of Neptune's moons.

Since Pluto has lost its classification as a primary planet, Neptune is now our Solar System's most distant planet and, technically, both our coldest and hottest planet. Because Neptune is an ice giant, it does not contain a solid surface like a terrestrial planet. Speculation is that under the dense gasses that encase the planet, an icy mantle surrounds a rocky core that is approximately the size of Earth. The gravitational pull from this core holds these gasses close to it and shrouds it from view.

Venusians have yet to officially relinquish their claim that Venus is our hottest planet at only 462°C (863.6°F). As NASA selects its focus for studies within our solar system over the next 20 years, strong consideration should be given to Jupiter, Neptune, and Triton. Astronomers believe they will be in a good position for a joint mission in 2038.

Summary Look

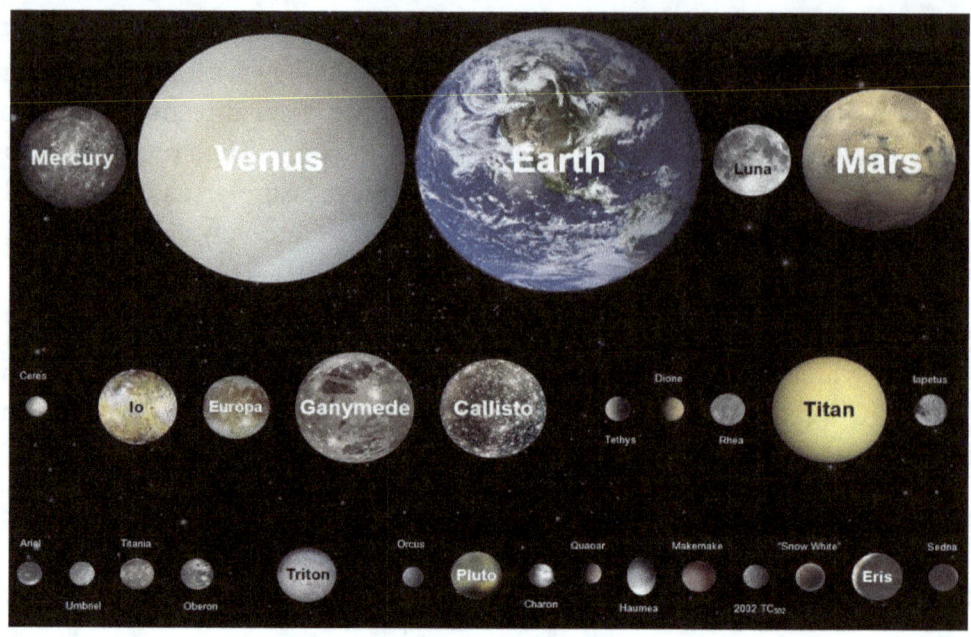

Credit: u/commander-crook
Planets and Various Moons in Our Solar System

We have now run out of planets in our solar system to explore physically or mentally.

But wait, there's still more!

Our Solar System is within Human's current capabilities

to explore for life — now! All it takes is vision,

courage, and commitment. However, this

turf represents only a relatively small

possibility of finding any form of

extraterrestrial life. If we

Humans are serious,

we have to go out

farther, much

farther out!

That is

to

Space, Ever Farther, Ever Faster – Now!

(next 2 pages high light exoplanets)

EXOPLANETS!

V HUMAN EXPLORATION EXPANDS

Chapter 22

Beyond Our Solar System to EXOPLANETS!

When we each think back to our early years and our first thoughts as our Sun slipped below the horizon and bright points of light appeared then multiplied and multiplied again and again to fill the heavens above us. We were often overwhelmed by their number, density, and closeness. With time and thought, we have come to understand that these bright points are stars, other suns, each with the possibility of planets surrounding them as we have around our own Sun—each one with the possibility of LIFE!

Like in our Solar System, life is not everywhere, but we cannot believe that we are a unique exception, and that life cannot exist elsewhere. The more other planets we find, the greater are the chances that we will also encounter extraterrestrial life.

So, with but a glance into the dark sky after sunset and the knowledge of our steadily increasing spaceflight capabilities, we become filled with anticipation, eagerness, and pride that we have arrived at this threshold. Our Moon was once at our threshold, but it is now behind us, replaced by other planets and moons in our solar system. But even before we have visited and personally explored any of these planets, exoplanets outside of our solar system (trillions and trillions of them) now offer even greater pulls on our psyche even though they come with much greater difficulties of someday personally exploring.

Could there be someone like one of us, or not too far from one of us, living light-years from Earth? US everywhere?

Those unlike us would be the most interesting!

The assumption here is a big one and an unjustifiable one unless we redefine "US". Humans, as we define ourselves, have evolved over several billions of years in response to the environment in which we have been "placed." We have no right to expect that anything we encounter would be like us. Living things, where they exist, would certainly not resemble US. Would they be highly inquisitive or only responsive to surrounding stimuli, extremely intelligent or dim-witted, huge or small, strong or weak, fast or slow, alive for centuries or only days, peace-loving or warlike, five senses like us or more or less, or.... we have no way to know until we encounter a fair sample of extraterrestrial life.

We have just begun the transition to seriously contemplating "Are we alone in this universe?" as a self-centered, theoretical, and philosophical one to one demanding real discovery, real scientific analyses, and a possible real realignment of our perceived role in our ever-expanding knowable universe. Changes forced upon us by new discoveries will, most likely, come on a generational time scale, not one demanding immediate year-by-year self-reassessments. But they will come!

So far, everything in our world is in place, and no discoveries across our solar system have changed that. But as we observe exoplanets from afar, starting with the 5,000 or so we have identified initially, we have every right to expect that the nature of worlds we encounter could be radically different than what we have encounter in the little corner of the universe to which we have been restricted to date.

"Are we alone?" So far, we have just eked the tip of our collective toe-off of our birth planet and can only respond as usual, "Yep. Haven't run into anybody else yet!" But the day is approaching when we will respond to the above question with a firm "NO!"

Enough with generalities and speculation! What do we know?

Currently, we know of approximately 5,000 extrasolar planets and climbing, but only a very small fraction of these are potentially habitable. Habitability depends on many factors such as the planet's atmospheric and surface temperatures, composition, gravity, and radiation environment. Only 12 planets, out of the approximately 5,000 "evaluated", are considered acceptable to support human life. Thus, finding another exoplanet that could support fragile beings like us or similar to us draws much attention.

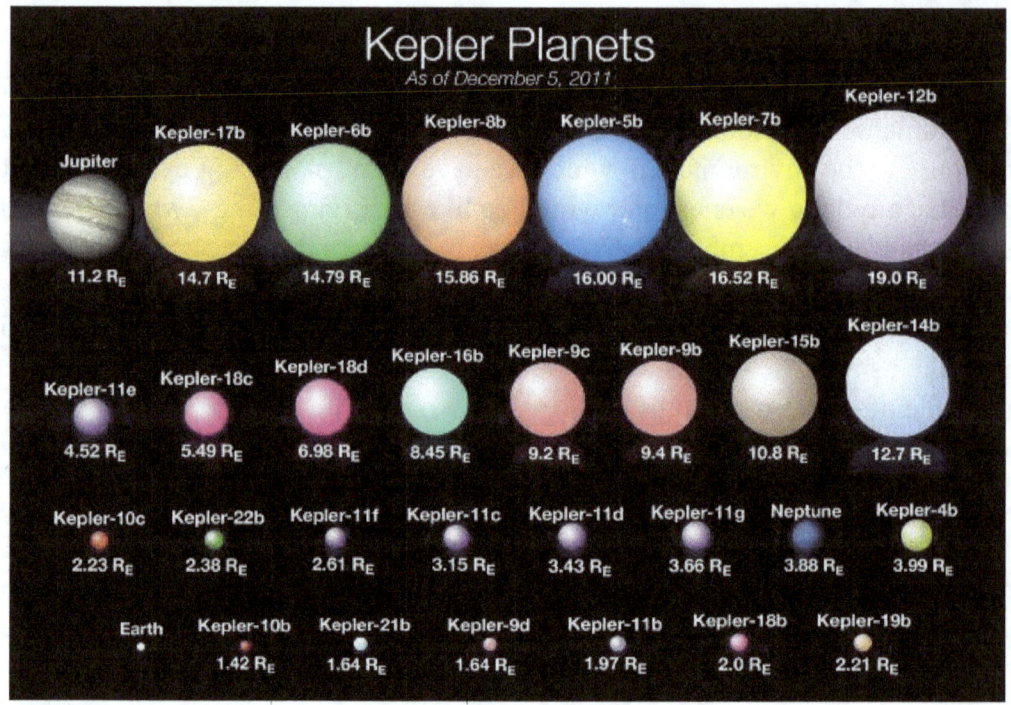

Credit: NASA/Kepler Mission
Earth plus 29 exoplanets identified by the Kepler Mission

The principle of planet detection is simple, but the telescope's operation is innovative technology. When a planet orbiting a star goes in front of that star, the observed light momentarily dips. From the frequency and strength of the dip, helpful information about the planet can be inferred.

A very broad-brush summary is:

1. Planets outnumber stars.
2. Small planets are common.
3. Both planets and solar systems are incredibly diverse.

Space, Ever Farther, Ever Faster – Now!

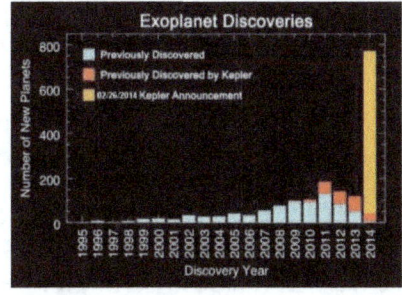

Credit: Wikipedia, The Free Encyclopedia

In 2013, Kepler ran out of fuel to repoint, but other methods were developed. By May 2016, Kepler had identified 1,284 new planets, of which 550 could potentially be rocky, but only nine of these orbits in their star's habitable zone. Over its 9-year lifetime, it had observed 530,506 stars and discovered 2,662 exoplanets.

Clearly, there has been a Sharp Rise in the Rate of Exoplanet Discoveries. Now the number is over 5,000 and climbing.

As of May 2020, TESS has identified 1835 exoplanets candidates; 46 have been confirmed as of July 2020. The launch, operation, and application to studying these and other candidates by the JWST are highly anticipated!

Fortunately, America was ready with the Transiting Exoplanet Survey Satellite (TESS) launched on a SpaceX Falcon 9 Booster in early 2018 into a highly elliptical orbit. It plans to study a sky area four hundred times larger than that covered by Kepler. The focus for the first two years will be on searching for transiting planets around candidate brighter, main sequence stars near Earth.

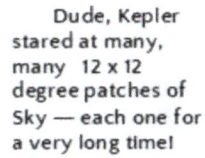

As of October 2019, TESS scientists have been collaborating with scientists from "Breakthrough Listen," dedicated to detecting any signs of advanced civilization in the data obtained. Though an exciting prospect, it is recognized that the probability of detection and recognition of such signals has an exceptionally low probability. Communication signals from remote, aging Captain Kirks or Spocks are even far less probable. However, if this effort continues and develops over a more extensive time, it might bear a little modicum yet exciting fruit.

"Meaningful two-way communication" with an intelligence external to Earth is almost completely out of the question using current technology. First off, the years between sending a signal and receiving a response are at least twice the distance away measured in light-years. Even more remote from possibility is one of us traveling out and back at the speed of light in the future. For that, we will need Einstein 2.0. Let us just stick to detecting candidate Earths outside our Solar System and scouring the associated data. TESS is in its infancy. Despite the pessimism, exciting progress in its development and returned data is upon us. Our ever-increasing space explorations, manned and unmanned, ever increase our chances of encountering some form of life out there, which would likely be far different from ours.

The chart at the beginning of the next section (Section V) has been assembled using the best of all available data on exoplanets "close" to us and in their stars' habitable zones. Each exoplanet has been partially brought to life through an artist's interpretation of the data acquired on them.

Chapter 23
Vision, Courage, and Commitment
Where Are We Today?

Vision – In general, we desire to return to the Moon, plan further expansion into our solar system, apply knowledge gained from unmanned explorations, develop required technologies for our explorations and applications here on Earth and stay ahead of other nations' accomplishments and technology developments.

Artemis is here and moving forward — at last. Change is coming!

Vision – Artemis is as clearly and strongly stated as our Apollo vision and will crystalize as we continue to focus our efforts. (Please see the Artemis discussion at the end of Chapter 19.)

Courage – Yes, the courage is here in specific government and free enterprise sectors to again step up to our vision and what it demands. It is hoped that our vision and courage to return to the Moon and then go on to Mars will remain as strong as required to reach our goal.

Commitment – One way to evaluate commitment is by the results achieved. Recently, America has returned to a firm commitment to space with Artemis and Space X leading the way. The continuous erosion of NASA's percent of our federal budget that has taken place since 1992 (page 89) has been halted and reversed!

A bright area in this picture is the accomplishments of America's free enterprise system led by SpaceX, Blue Origin, Boeing, Bigelow Aerospace, and others. The cost of space flight has been greatly reduced while the commitment, flexibility, and agility of their associated space efforts remain high.

A second bright area was made at the presidential level (2017-2021), which did not settle for the inadequate performance (summarized above) and as was enumerated by Vice President Mike Pence, who in 2019 stated that:

1. "We are racing against our worst enemy: complacency.
2. We will not settle for anything less than boots on the Moon by the end of 2024.
3. NASA must transform itself into a leaner, more accountable and more agile agency. Moon in next five years, then commercial rockets it will be."
4. Failure to land in five years is not an option, and this specific landing date has the sense of purpose and urgency.
5. It is apparent that major changes are called for within NASA to decrease the time to deliver results and lower costs."

We were and are in a 21st-century space race with China and Russia. The United States must remain first in space in this century as in the last, not just to propel our economy and secure our nation but, above all, because the rules and values of space, like every great frontier, will be written by those who have the courage to get there first and the commitment to stay.

It is highly admirable and uplifting that the Trump administration has taken the interest, developed the knowledge of our space program, and is so adamant about timely performance and lower cost. However, unlike the days of JFK and Apollo, today we have a $37 trillion national debt, and most of the major support required comes from the government side out of our highly restricted discretionary spending. (Please see "American Private Enterprise is Moving In" on pages 236 - 237.)

Fortunately, roles are slowly evolving, government leadership is shifting into government regulation, and commercial contracting is shifting into free enterprise leadership. That is, our space industry is gradually taking on more of the functional appearance of our aviation industry.

Several previous administrations have also produced strong rhetoric favoring our space effort but failed to produce the desired results. However, the growing strength of our free-enterprise commercial space companies is now able to step in where the federal government is unable to perform. With our government's traditional space contractors and relatively new contractors working as a team, the results we have sought for many years may now be upon us.

Space, Ever Farther, Ever Faster – Now!

With this new, strong, and consequential vision, courage, and commitment as We, the Human, Enters Space, advancements in space exploration will be coming more rapidly.

The words and actions of the President Trump (2017-2021) instilled a new vigor in our human space effort and give us hope for a return to a program functioning on all cylinders. As Vice President Pence has stated, "The agency required just eight years to go to the Moon at a time when NASA did not know how to do the job. Now NASA has said that it cannot land humans on the Moon before 2028, more than 11 years after President Trump first established the goal of returning humans to the lunar surface. Ladies and gentlemen, THAT is just not good enough. We are better than that."

This new program and the strong leadership behind it are renewing hope and excitement in the minds and hearts of individuals who were part of NASA in the 1960s and those of today who are on the front lines making it happen.

But first, we must enter the realism of politics that opposes cost, schedule, or performance improvements. Some current plans must be reconsidered:

1. The expensive Lunar Gateway must be cut back or even eliminated.
2. The use of entrepreneurial commercial components, individual systems, and total system replacements must be considered if cost, schedule, or performance are more advantageous.

These actions may not be easy. If SpaceX takes over an existing NASA project based on cost and schedule considerations, it will not be popular in some Congressional delegations and might work against other NASA funding and progress.

> **Our initial drive into space yielded MAJOR benefits for America.**
>
> *Can free enterprise, leadership and we make it happen again — NOW*

Yep, you bet it can! But you guys sure screwed it up!

Yeah, we got it now. You guys go to the corner!

Right Guys, just GO, SIT, STAY!

***

Chapter 24. Our Drive to Explore

As stated on the first page of this book:

"The essence of our species is to explore — to find new answers and new meaning for who we are."

—Pat McCarthy, Vice President, GM

Human Exploration Cannot Be Stopped!

`Our drive to explore is irrepressible, a part of our human spirit, and an inherent part of our DNA. Think back to just a couple hundred thousand years, to our distant ancestor, Gork, a really good guy, and his trusty sidekick, Grog.

Well, today is the day. They cannot put it off any longer. They have to get over that first hill, then many more to ultimately reach the fetching top of that mountain. With determination, they head out despite intense pressure from the little women to stay home and clean their caves.

Credit: Burian, "Prehistoric Man"

Neanderthals had our drives but with a bit less technology.

They felt inhibited by the first stream but struggled across it, then conquered more hills, streams, and rivers until they reached the foot of that giant mountain.

The Prize!

With time, they struggled to the top — just to get a better view of the territory around them, just to get to that beautiful peak to discover more about the world in which they had found themselves.

Then again…

Eventually, some of their distant descendants with that same spirit of exploration headed out over the ocean towards that far-away, mysterious horizon despite intense pressure to stay home and not fall off the edge of the Earth.

Space, Ever Farther, Ever Faster – Now!

It was just a little over that inviting horizon…

Credit and much appreciation to Ed Miracle, artist

Now, We Humans, with our improved eyes (the Hubble Space Telescope and now the James Webb Space Telescope) and our improved transportation (SpaceX and Artemis) the door is wide open to finding new life in new planetary systems!

Although there will be a continuous turnover of enthusiastic individual participants, WE, THE HUMAN, continues outward from Earth unabated.

Chapter 25
Where Does Human Go Now?

Human has much to think about, to ponder … Maybe Human should let its conscious also have an input:

Human — Who's there?

Conscious – ME!

Human — Oh, YOU again, Conscious. What do you want THIS time?

Conscious – First of all, will you stop being so formal! Just call me Connie.

Human — OK Connie, whatever you want.

Conscious – Good! Now, are you finally going to take your foot off that doorstep and go out to someplace… just any place?!

Human — Nag, nag, nag. Don't rush me!

Conscious – Rush you? You have been hanging around down here, barely getting off your duff for about the past 200,000 years!

Human — Well, I did leave Earth orbit SEVEN TIMES just SIX decades ago!

Conscious – Just a tiny, tiny shuffle, a blip in time!

Human — Well, it's a start…

Conscious – Start? All you did was go to the nearest moon you could find!

Human — Gotta start somewhere…

Conscious – What about all the other moons in just our solar system, then those beyond?

Human — Beyond? THAT would be really hard to do now!

Conscious – What a wimp! Think of those rewards we'll get… like all the new species we will encounter and all the new technologies we will develop!

Human   – Yeah, yeah, that would be interesting…

Conscious – NO! Not just interesting, but in our future, we might need what you learn for our own survival.

Human   – No sweat, that's a long way off.

Conscious – What about Mars that's coming right up!

Human   – Ahh, well ahhh… ahhh…

Conscious – Complacency kills...

Human   – You are right there, Connie. I'll watch out.

*Conscious – Good. **YOU make sure you DO!***

Human   – Yes, Mam…

*Conscious – Remember, all it takes is **Vision, Courage, and Commitment!***

APPENDIX
Abbreviations

AAP Apollo Applications Program

ACTS Advanced Communications Technology Satellite

AF Air Force

AFB Air Force Base

ALSEP Apollo Lunar Surface Experiments Package

ALT Approach and Landing Test

ATK Alliant Techsystems Inc.

ATM Apollo Telescope Mount

BAE British Aerospace

CIT California Institute of Technology

Capcom Capsule Communicator

CASIS Center for the Advancement of Science in Space

EVA Extra-Vehicular Activity

FBI Federal Bureau of Investigation

FEMA Federal Emergency Management Agency

HST Hubble Space Telescope

ISS International Space Station

JPL Jet Propulsion Laboratory

JSC Johnson Space Center

JWST James Webb Space Telescope

KSC Kennedy Space Center

LBJ Lyndon Baines Johnson

LC Launch Complexes

LM Lunar Module

LMP Lunar Module Pilot

LRV Lunar Roving Vehicle

MIT Massachusetts Institute of Technology

MET Modular Equipment Transporter

MSFC Marshall Space Flight Center

Ph.D. Doctor of Philosophy

RCS Reaction Control System

SIM bay Scientific Instrument Module Bay

SM Service Module

SPACE Space Policy Advancing Commercial Enterprise

SpaceX Space Exploration

SPHEREx Spectro Photometer for the History of the Universe, Epoch of Reionization and Ices Explorer

SPS Service Propulsion System

SRB Solid Rocket Booster

STEM Science, Technology, Engineering, and Math

STS Shuttle Transportation System

TDRS Tracking and Data Relay Satellite

TSS Tethered Satellite System

ULA United Launch Alliance

USS United States Ship

UV Ultraviolet radiation with wavelengths shorter than visible light and longer than X-rays

UVIS Ultraviolet Imaging Spectrograph

VAB Vertical Assembly Building

WFC Wide Field Camera

X-rays Electromagnetic radiation with wavelengths shorter than ultraviolet rays and longer than gamma rays

Book References

1. Emily Carney - https://space.nss.org/space-myths-busted-no-there-wasnt-a-mutiny-on-skylab/
2. "Fifty Years of Research in Helium-3 Fusion and Helium-3 Resources", J.C. Crabb, S.W. White, L.P. Wainwright, S.E. Kratz, G.L. Kulcinski, Fusion Technology Institute University of Wisconsin-Madison, UWFDM-935, http://fti.neep.wisc.edu
3. "Return to the Moon," Harrison H. Schmitt, Copernicus Books, Praxis Publishing, 2006. Many of the issues discussed here on Helium-3 and our future lunar program have been provided by personal contact with Harrison Schmitt and from this excellent, expansive yet comprehensive publication.
4. In coordination with Harrison Schmitt, private communications
5. Gibson, E. G., The Quiet Sun, NASA SP-303. 1973 pp 82 & 107
6. https://en.wikipedia.org/wiki/Lunar_Orbital_Platform-Gateway
7. Berger, Eric. "Former NASA administrator says Lunar Gateway is "a stupid architecture". Ars Technica. Retrieved 23 November 2018.
8. Foust, Jeff. "Advisory group skeptical of NASA lunar exploration plans". Ars Technica. Retrieved 20 December 2018.
9. "Op-ed: The Deep Space Gateway would shackle human exploration, not enable it". Ars Technica. Retrieved May 20, 2018.
10. Moon Direct, The New Atlantis, Number 56, Summer/Fall 2018, pp. 14-47.
11. The Lunar Orbital Platform-Gateway: an unneeded and costly diversion. Gerald Black, The Space Review. 14 May 2018.
12. Whittington, Mark. "NASA's unnecessary $504 million lunar orbit project doesn't help us get back to the Moon". The Hill. Retrieved 20 December 2018.
13. Berger, Eric, "Chinese Space Official Seems Unimpressed with NASA's Lunar Gateway." ARS Technica. Retrieved 17 July 2018
14. Health threat from cosmic rays – Wikipedia; Figure: "Comparison of radiation doses."
15. https://en.wikipedia.org//Health_threat_from_cosmic_rays
16. Some data extracted in part from: "China is beating the United States in the New Space Race." Op-ed by Brandon J. Weichert, Space News, 2/1/19 https://spacenews.com/op-ed-china-is-beating-the-united-states-in-the-new-space-race/

17. https://en.wikipedia.org/wiki/Titan_(moon)

Skylab References

18. Alexandra and Dwight-Steven Boniecki, Searching for Skylab - https://searchingforskylab.com
19. Bakers, David, NASA Skylab 1969-1979 (all models); Owners' Workshop Manual, Haynes Publishing, 2018
20. Belew, Leland F. (editor) Skylab, Our First Space Station, NASA SP-400, 1977
21. Belew, Leland F. and Stuhlinger, Ernst Skylab, A Guide Book, NASA EP-107, 1973, GSFC
22. Compton, W. David, and Benson, Charles D.: 1983, Living and Working in Space: A History of Skylab, Washington, DC: NASA SP-4208, pp. 379-386
23. Cooper, Henry S.F. Jr. A House in Space, Holt, Rinehart, and Winston, 1976
24. Eddy, John A. A New Sun, The Solar Results From Skylab, NASA SP-402, 1979
25. Elder, Donald C.: 1998, The Human Touch: The History of the Skylab Program, Chapter 9, From Engineering Science to Big Science, The NACA and NASA Collier Trophy Research Project Winners, Ed. Pamela E. Mack, The NASA History Series, NASA Office of Policy and Plans, NASA History Office, Washington, D.C.
26. Hitt, David; Garriott, Owen and Kerwin, Joseph, Homesteading Space, Nebraska Press, 2008
27. Pogue, William R.: 2011, But for the Grace of God, Soar with Eagles, Rogers, Arkansas
28. Shayler, David J.: 1984, Around the World in 84 Days, An Authorized Biography of Skylab Astronaut Jerry Carr, An Apogee Books Publication, Ontario, Canada
29. ★ Shayler, David J.: 2001, Skylab, America's Space Station, Springer-Praxis series astronomy and space sciences (AN EXCELLENT BOOK)
30. ★ Shayler, David J.: 2020, America's First Space Shuttle Astronaut Selection, Springer-Praxis Books in Space Exploration
31. Steven-Boniecki, Dwight: 2015, Skylab 1 & 2, NASA Mission Reports, Apogee Books Space Series

32. Steven-Boniecki, Dwight: 2016, Skylab 3, NASA Mission Reports, Apogee Books Space Series, Steven-Boniecki, Dwight: 2016, Skylab 4, NASA Mission Reports, Apogee Books Space Series
33. ★ Steven-Boniecki, Dwight: 2019, Searching for Skylab (an Outstanding Oscar-eligible film), 1080VMC & AHAB Productions, directed by Dwight Steven-Boniecki, a film in which my daughter and I were privileged to appear.
34. Stuhlinger, Ernst: 1970, The Skylab Story, Astrodigital, Space, Skylab
35. Summerlin, Lee B. (editor) Skylab Classroom in Space, NASA SP-401, 1977
36. Biomedical Results from Skylab, NASA SP-377, 1977
37. Skylab Astronomy and Space Sciences, NASA SP-404, 1979
38. Skylab Chronology, NASA SP-4011, 1977
39. Skylab Explores the Earth, NASA SP-380, 1977
40. Skylab EREP Investigations Summary, NASA SP-399, 1978

Further Considerations on The Lunar Gateway

The Gateway does have the advantages of providing missions for the expensive SLS and is something that we know how to do. However, let's do what is more challenging and will yield significant results quickly, efficiently, at lower cost, and continuously.

The Gateway is stated to have the additional benefits of advanced research and technology development, expansion of our space economy, enabling international and commercial participation and leading to a sustained lunar presence, which are all worthwhile benefits. However, corresponding developments directly on the lunar surface will yield equivalent benefits, if not more. More than enough high return space projects are calling to us; we do not need to invent and tie ourselves down to other projects that lack equivalent returns but consume significant opportunity and funding.

As stated in Wikipedia (5), many other experienced space professionals have voiced their opposition to the Gateway, including:

1. Michael Griffin, a former NASA administrator, stated that "putting a Gateway before boots on the Moon is, from a space-systems engineer's standpoint, a stupid architecture." (6)
2. Former NASA Astronauts Eileen Collins, who was a Space Shuttle pilot and commander, and Harrison Schmitt, who was Lunar Module pilot aboard Apollo 17, did not mention the Gateway specifically but criticized NASA's plans for lunar exploration for not being ambitious enough; that is, the slow pace of getting back down on the lunar surface. (7)
3. Former Apollo 11 astronaut Buzz Aldrin stated that he is "quite opposed to the Gateway and that using it as a staging area for robotic or human missions to the lunar surface is absurd." Aldrin also questioned, "Why would you want to send a crew to an intermediate point in space, pick up a lander there and go down?" Instead, Aldrin expressed support for Robert Zubrin's Moon Direct concept, which involves lunar landers traveling from Earth's orbit to the lunar surface directly and back. (8)
4. Former NASA Astronaut Terry Virts, a pilot of STS-130 aboard Space Shuttle Endeavour and Commander of the ISS Expedition 43, stated, "Gateway would shackle human exploration, not enable it." He cannot envision a new technology that would be developed or validated by

building another modular space station and criticized NASA for abandoning its safety dictum of separating the crew from the cargo at launch that was established following the Space Shuttle Columbia accident in 2003. (8)

5. Mars Society founder Robert Zubrin, who has been a strong supporter of a human mission to Mars, stated that "The Gateway is NASA's worst plan yet. It is just a toll booth. We do not need a lunar-orbiting station to go to the Moon or near-Earth asteroids. We do not need such a station to go to Mars or anywhere. It does nothing that we can't do in the ISS, and there is nothing at all in lunar orbit: nothing to use, nothing to explore, nothing to do. If the goal is to build a Moon base, it should be built on the surface of the Moon. That is where the science is, that is where the shielding material is, and that is where the resources to make propellant and other useful things are to be found." (9)

6. Retired aerospace engineer Gerald Black stated that the Gateway is useless for supporting the human return to the lunar surface and a lunar base. It is not planned to be used as a rocket fuel depot or stopping there on the way to or from the Moon would, which would only waste rocket fuel. (10)

7. A contributor to the Hill Newspaper and author of several space exploration studies, Mark Whittington, stated that "NASA's unnecessary $504 million lunar orbit project does not help us get back to the Moon. It was not necessary for Apollo. Also, a reusable lunar lander could be refueled from a depot on the lunar surface and left in a parking orbit between missions without needing a big, complex space station." (11)

8. Pei Zhaoyu, deputy director of the Lunar Exploration and Space Program Center of the China National Space Administration, concludes that, from a cost-benefit standpoint, the Gateway would have lost cost-effectiveness and that the *Chinese plan is to focus on a research station ON THE SURFACE*. (12)

9. Robert Zubrin has made a detailed comparison of the Gateway, and Moon Direct approaches. A few of his numerous observations are:

"Timing of Return to Earth orbit - Return to Earth orbit from Gateway's lunar orbit must be precisely timed. In contrast, direct return from the lunar surface does not require precise timing because the Earth is always at the same point in the lunar sky. The Moon's rotation is tidally locked to the Earth, so we always see the same side of the Moon from Earth. Direct return to Earth orbit is always available.

"Rocket System Requirements – Gateway plans to use the SLS booster and the Orion spacecraft, which are not yet operational and very expensive. Moon Direct can all be done with rockets already commercially available; that is, SpaceX's Falcon 9 and Falcon Heavy. Once lunar propellant production is online, each recurring mission could be done through a single Falcon 9 launch. Only one Falcon 9 is needed to deliver a new crew to lunar orbit in a Dragon 2 spacecraft, exchange crews in a waiting Lunar Excursion Vehicle (LEV) in lunar orbit, and refuel it. A new crew would fly to the Moon in the LEV, which refuels at the lunar base while the Dragon 2 returns to Earth with the previous crew.

"Cost - The SLS has been projected to cost over $2 billion per year for the five years starting in 2018. Also, the use of the SLS to launch both equipment and crews requires that it be the first man-rated, which introduces considerable additional cost. NASA's total human spaceflight program total budget is currently around $10 billion per year.

"In contrast, because Moon Direct requires relatively little launch mass, largely uses existing technologies, and can be implemented much more cheaply than the Gateway, at the cost of approximately $420 million per year.

"SpaceX's Falcon 9 with a full payload has approximately $70 million likely launch cost. If we include the cost of the propellant, cargo, and crew, the total cost of each recurring mission would be roughly double the launch cost, or about $140 million — low enough for a highly sustainable lunar exploration program. Once lunar-produced propellant is available at a site, the mass and expense of recurring lunar missions will drop dramatically, as will the cost of recurring flights across the surface. A Falcon Heavy likely launch cost is $150 million and will increase similar to that of the Falcon 9 cost as stated above.

"Efficiency – In the Moon Direct mode, crews will live and work continuously on the lunar surface, not way above in an orbiting gateway with only sporadic tours of duty on the surface. They will immediately get to work mining propellant and performing the first sustained, global human Moon explorations.

"Launch Mass Requirements and Additional Considerations [Zubrin continued – (9)]

Initial Mass in Low Earth orbit per Mission – tons		
Phase	**Gateway**	**Moon Direct**
1. Unmanned missions deliver materials to Moon for lunar base and propellant production	520	120
2. Two Piloted missions make base operational, including propellant production	230	224
3. Twenty recurring piloted missions use propellant for exploration and return to Earth orbit	2,000	300
Total	2,750	532
Additional Considerations		
1. Percent of lunar surface available per mission	3	25
2. Complications and hazards of lunar orbit rendezvous	Yes	No
3. Sources of required cosmic radiation shielding protection	Must be brought up from Earth	Lunar soils and water, lava tubes, and Moon itself

"Thus, even though the Gateway requires over five times the mass to be delivered to low Earth orbit, the Moon Direct opens up over eight times of the lunar surface for exploration."

Granted, both the Lunar Gateway and the Moon Direct require the construction of stations; however, in Moon Direct, the station, crew, and all necessary equipment are where it counts—on the surface to be explored, not high above in an orbiting Gateway space station.

NASA is currently struggling under the financial burden of one space station: why build another?

It is apparent that there are strong reasons not to support the current NASA approach to lunar exploration. However, NASA appears to be off and running and will not be deterred unless instructions come from above. It is also apparent that a few entrepreneurial commercial firms and some internationals, China, in particular, have the motivation and capability, perhaps even the necessary funding, to go directly to the lunar surface, arrive before NASA, and be more efficient and productive.

Perhaps the motivation to reconsider the Gateway will come via the firm request to NASA from President Trump and Vice President Pence (March 2019) to put humans on the Moon by 2024. Since the Gateway, as currently envisioned, will not be completed by 2026, NASA has a decision to make at its very best.

The Gateway is not a rational way to proceed. For We, the Human, Enters Space and NASA, I hope the Gateway will soon be reconsidered. However, some already onboard contractors will clearly financially benefit, and other nations that join in will have an additional focus for their work and support. Will committed money win out over reason?

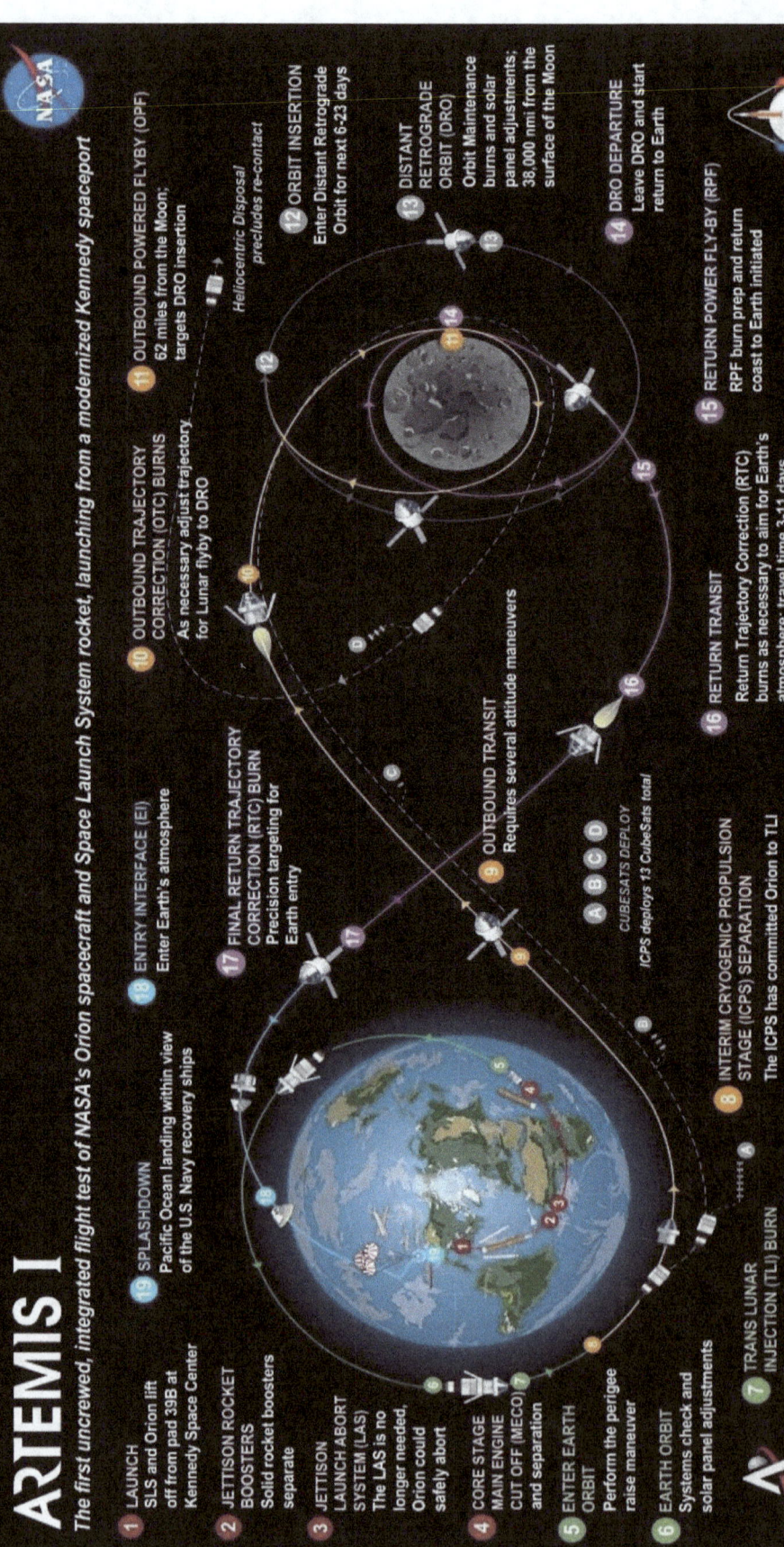

Ed Gibson

Ed Gibson Resume – Brief Summary

Ed earned a BS in Engineering from the University of Rochester (1959), an MS (1961), and a Ph.D. (1965) in engineering and physics from the California Institute of Technology.

In 1966, Ed was one of six scientists selected to the first group of Scientist Astronauts out of 1,100 applicants. He earned Air Force wings (1967) at Williams AFB where he graduated second in in his class. He has logged over 2,200 hours in high-performance aircraft, 100 hours in helicopters, and 2017 hours in space during his career.

Ed supported the Apollo 12 crew by helping design their spacewalks and being their ground communicator (Capcom) during their three (Extra Vehicle Activity (EVA) explorations of the moon.

In 1973 he and his two crewmates lifted off, rendezvoused with the Skylab Space Station, and remained there for 84 days, which lasted as an American record for 21 years. On that last mission of Skylab, Ed worked outside the space station for 15 hours during three different spacewalks.

After his NASA career, Ed entered program management and marketing with Booz, Allen, and Hamilton and with TRW, where he focused on space and energy development. He later served as the President of the Oregon Museum of Science and Industry and his own consulting company, Gibson International. He retired as a Senior Vice President with Science Applications International Corporation and as one of two Managers at Aerospace Partners, LLC, and as Chairman of a NASA Independent Review board for America's return to the moon.

Ed has also published many technical papers, one textbook (The Quiet Sun, a NASA Publication), many scientific articles for scientific audiences and the general public, and two novels (Reach and In the Wrong Hands) published by Doubleday, Bantam, and the MacDonald Co. in London, Great Britain.

Now Ed is publishing this 300-page book "Space-Ever Farther, Ever Faster-Now" in which he described America's motivations and each individual mission within Mercury, Gemini, Apollo, Skylab, Apollo-Soyuz, Space Shuttle, International Space Station as well as our return to the Moon and Landings on Mars, and farther out.

In 2021, Ed had asteroid 132603 named after him.

Ed Gibson Resume – Briefer Summary

Spacewalk Training
WATER TANK – 1975

Reflection Time
SWIMMING POOL – 2024

We All Are Fortunate!

"Like just about every citizen of America and other freedom-loving nations, I have been fortunate. I pursued a career in space exploration without any major barriers except for those supplied by mother nature or I constructed for myself. Very early in my life, my future career could not have become even a dream.

When I was but 2 years old, I contracted osteomyelitis in my right shin bone, a bacterial infection that results in a softening of the bone to the point that the weight of even a toddler could not be supported. In the 1920s and early 1930s, if the infection could not have been removed with a scalpel, it would have been removed with a saw along with a leg or arm.

However, I was fortunate. In London on September 28, 1928, Dr. Alexander Fleming announced the discovery of penicillin, which was then developed over several years for practical and safe medical application. With a few years to spare, it was ready to save my leg in 1938, then in again in 1940 and 1946 when further developments of penicillin were available to cure additional recurrences of the infection.

I suspected that NASA did not want any one-legged astronauts. However, by the time I applied for the first Scientist Astronaut selection in 1965, NASA judged that since the infection had not reoccurred in 19 years, the bacteria in my right shin bone was dead and I was cleared to go through the selection process. Unfortunately, one other candidate with a much later occurrence of osteomyelitis, was not cleared.

Yes, I was fortunate!

However, I just about lost it all because of my less-than-stellar academic performance. I failed first grade, then demonstrating strong consistency, I failed fourth grade as well. Granted, I was in hospital beds and wheelchairs part of the time, but poor grades were just that: poor grades.

As I approached high school, I was forcefully and continuously informed that hard work improves performance.

'What?! Really? Who knew?!'

I began to work harder, and soon my grades skyrocketed all the way up to mediocre. My grades continued to improve through high school, so much, in fact, that before graduation, I was accepted for admission to the University of Rochester, a top-quality academic institution 60 miles up the road from our home in Kenmore, New York. It made my day — and my life!

Yes, I was in America where every day is a new day and current performance can overshadow past performance.

For sure, I was fortunate!

Once in college, hard work became a way of life. At the end of the first term, one of my friends said that he had seen my name on the Dean's List. 'What? I never even met this Dean guy. What did I ever do to him? Why am I on his list?'

Before graduation I was awarded the future financial support in the form of a National Science Fellowship along with acceptance to the California Institute of Technology (Caltech) in Pasadena, California to work towards an MS and PhD in physics, jet propulsion and rocketry.

Clearly, although my performance had started at zero, America had faithfully supplied the freedom and opportunity to improve and achieve.

I was also fortunate to also receive strong and consistent support from my parents first, then from my inside-and-outside beautiful girl friend, Julie, who since 1959, has not only been my best friend but also my wife.

How fortunate can one guy be!"

Focus on Students

"One of my intents in writing this book has been not only to get you, the reader and you, the student, more interested in space but also to provide you with the understanding and confidence that enable you to also participate and contribute as many others have done in the past and to the best of your ability.

I know that if, when in Elementary School, I was told that someday I might be able to do what is summarized in my resume, I would have smiled, run, and found cover in some distant bushes. Yet, just about every student reading these words has more than enough ability to get the job done with excellence and take pride in their performance. Speaking from the vantage point of the kid out there hiding in the bushes, here are some of the things I eventually learned:

"Never sell yourself short!" That is, never assume that your future accomplishments are limited to only what you have achieved in the past!

My academic record started out on the wrong end of accomplishment. But eventually I was fortunate to be accepted into the engineering program at the University of Rochester (UR), an excellent school!

I had also applied to Cornell, but they didn't want me in or near their university (slight exaggeration); however, after college, they offered me a very generous fellowship when I applied to their graduate program. For sure, it is true, as I was about to learn again, hard work does pay off.

When I had arrived at the UR, I clearly understood that it was "NOW OR NEVER!"

Success is also enabled by support from someone close to you. When I received a Ph.D. from Caltech, my wife, Julie, received a Ph.T. (Put husband Through).

Success is also more likely if you recognize your strong points.

Finally, I realized that I was good at subjects that predominately used logic and reasoning (math, physics) rather than memory (history, languages, and organic chemistry). I used to think I was good at history and languages, but then… maybe not… I don't recall.

Success breeds more hard work and more success. The rest is in my resume.

So, what is OUR conclusion?

If this road apple kid could achieve his dream through hard work, SO CAN YOU!

Success in academics, as well as most other pursuits in America and other freedom-loving nations, is proportional to the effort that you put into pursuing your goals. There are no surprises here.

Therefore,

<u>*UNDERSTAND AND FULLY BELIEVE THAT*</u>

<u>*SUCCESS IS WITHIN EACH ONE OF YOU!*</u>

GO FOR IT, and NEVER, NEVER, <u>NEVER GIVE UP!</u>"

Note to Author…

"Hey, you, Smarty Pants.

Think you got it all covered…

our Universe?

Well…… you completely left out the

Multiverse!"

EXOPLANETS!

Ed Gibson

www.ingramcontent.com/pod-product-compliance
Lightning Source LLC
Chambersburg PA
CBHW081718100526
44591CB00016B/2418